BEZ6251

Temperature Regulation in Humans and Other Mammals

Springer

Berlin
Heidelberg
New York
Barcelona
Hong Kong
London
Milan
Paris
Singapore
Tokyo

Claus Jessen

Temperature Regulation in Humans and Other Mammals

With 100 Figures

 Springer

Professor Dr. Claus Jessen
Justus-Liebig-Universität Gießen
Physiologisches Institut
Aulweg 129
35392 Gießen
Germany
and
Redwitzstr. 5A
96191 Viereth-Trunstadt
Germany
E-mail: Claus.Jessen@t-online.de

ISBN 3-540-41234-4   Springer-Verlag Berlin Heidelberg New York

Library of Congress Cataloging-in-Publication Data
Jessen, Claus, 1935-
Temperature regulation in humans and other mammals / Claus Jessen
p. cm. Includes bibliographical references and index.
ISBN 3540412344 (alk. paper)
1. Body temperature--Regulation. I. Title.
QP135 .J47 2000   571.7'619--dc21   00-049708

Springer-Verlag Berlin Heidelberg New York
a member of BertelsmannSpringer Science+Business Media GmbH

© Springer-Verlag Berlin Heidelberg 2001
Printed in Germany

Cover Design: *design & production*, Heidelberg
Typesetting: Camera-ready by the author

SPIN: 10771946   31/3130xz – 5 4 3 2 1 0 – Printed on acid free paper

Preface

In 1996, environmental physiology was the subject of two volumes published in the Handbook series of the American Physiological Society. The largest section deals with temperature regulation. It covers the research of the preceding 30 years and is an excellent reference book for advanced scientists working in this or closely related fields. However, handbooks are rarely read from cover to cover. This text is intended to fill the gap between detailed information for specialists and the more cursory treatment of temperature regulation in general textbooks of physiology. It is a brief comprehensive presentation, melding the results of recent research into the background knowledge collected in earlier periods. The intended audience for this book includes students, teachers, physicians and scientists interested in various aspects of thermal physiology, such as exercise, adaptation, circulation, fluid balance or physiological regulation in general. Any other reader is welcome to share my interest in this fascinating field of integrative physiology.

Viereth–Trunstadt, September 2000

Contents

1 Introduction

Temperature regulation is one of several regulatory systems which maintain important constituents of the body's internal milieu such as osmolality, gas pressures, blood pressure and temperature, at levels whose combination may be the best compromise between conflicting demands. This book concentrates on temperature, and may occasionally convey the impression that a constant body temperature is of paramount importance to survival. However, this is not the case. Animals in natural habitats often have to rank the responses to thermal stress in the context of other stresses which concurrently may be placed upon them. The task of the temperature–regulating system is to maintain the temperature of the body within a range compatible with the functions of other regulatory systems.

1.1
Poikilotherms and Homeotherms

The range is wide in poikilotherms such as insects, fish and reptiles; whose body temperature usually varies with and nearly to the same extent as the temperature of the environment. This does not exclude that some species prefer certain temperatures and use behavioural means of avoiding others. The point is that, in poikilotherms, widely different body temperatures do not critically interfere with vital body functions. As evidenced by their survival, this coarse type of regulation is sufficient, or may even be the superior strategy in particular ecological conditions.

Mammals and birds are homeotherms. In most placental mammals, the lifetime mean of internal body temperature is in the range of 36–40 °C, while the average temperature of the environment is usually much lower. Thus, a relatively large temperature gradient between the internal body and its environment is maintained throughout life. A second feature of homeothermy is that the internal body temperature of a standard mammal rarely leaves the range of ±2 °C around its lifetime mean, even if the temperature of the environment varies to a large extent. Homeothermy is characterized, in the face of a mostly colder and variable environment, by the **maintenance of a high body temperature within a narrow range**.

A living animal is producing heat by its metabolism, and normally losing heat to the environment. If the temperature of a homeotherm is to remain constant over time, **the rates of heat production and heat loss must be equal**. This is a case for Newton's law of cooling: **the rate of dry heat transfer** between an object and its environment is, in a first approximation, proportional to the temperature difference. Imagine an **inanimate object** of 37 °C temperature, in which heat is released from an electric heater at a constant rate of 70 W (Fig. 1.1). If the object is placed in an ambience of 37° C, the heat loss by radiation, convection and conduction is zero, and its temperature must rise because heat continues to be liberated. With decreasing ambient temperature, heat loss increases and may be assumed to reach 70 W at 28 °C. At this single temperature, the rates of internal heat release and heat loss to the environment are in **passive equilibrium**, and the temperature of the object remains constant over time. At ambient temperatures below 28 °C, however, the rate of heat loss is larger, and the object cools down.

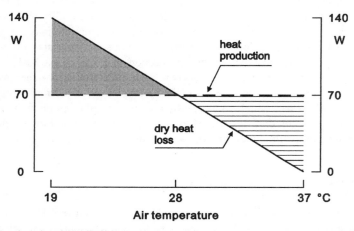

Fig. 1.1. Dry heat loss by radiation, conduction and convection (*solid line*) of an inanimate object of 37 °C temperature as a function of ambient temperature. In the object, heat is released from an electric heater at a constant rate of 70 W (*interrupted line*). *Grey and hatched areas* see text

Also, a **living animal** must obey Newton's law. If the temperature of a homeotherm is to remain constant, the **equilibrium** between the production and loss of heat must be actively maintained **under all environmental conditions**. It follows that, at ambient temperatures below the point of passive equilibrium, the regulating system must increase the rate of heat production, by whatever means, so that it matches the rate of heat loss (Fig. 1.1: grey triangle). To prevent body temperature from rising at ambient temperatures above the passive equilibrium, the deficit of dry heat loss must be made up by **evaporative mechanisms of heat loss** such as panting or sweating (Fig. 1.1: hatched triangle).

If the difference between homeotherms and poikilotherms is projected onto this background, two unique properties of homeotherms appear instrumental in defending a high body temperature against a cooler environment (Fig. 1.2). The first is **tachymetabolism**. At the same body temperature, the resting metabolic rate or heat production of a **tachymetabolic mammal** is approximately three to four times larger than that of a **bradymetabolic reptile** of equal body mass [111,499]. The second is that homeotherms have developed means of overcoming the constraints of the Arrhenius–van't Hoff relationship. If a poikilotherm is subject to a 10 °C decrease in body temperature, its resting heat production is 60% lower than before, equivalent to a Q10 of 2.5 [503]. **In homeotherms**, however, **heat production increases with decreasing body temperature**. It is the combined result of both features – tachymetabolism and overcoming the Q10 effect by regulation – that the internal body temperature of a standard homeotherm in cool environments remains high and largely independent of ambient temperature.

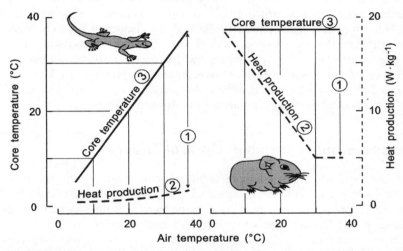

Fig. 1.2. Relationships between air temperature (*abscissa*), internal body temperature (core temperature *left ordinate*), and heat production (*right ordinate*) in a standard reptile (*left*) and a mammal (*right*) of equal body mass. **1** At 38° C body temperature, the heat production of the mammal is three to four times larger than that of the reptile. **2** With decreasing air temperature, heat production of the mammal increases, while it decreases in the reptile. **3** Consequently, the internal temperature of the mammal is largely independent of air temperature, while that of the reptile is a passive function of air temperature

1.2
The Capacity of the System

If the heat loss to the environment is larger than the maximum of heat production, the temperature–regulating system is overtaxed and body temperature must fall. The maximum heat production of a medium–sized species as a response to cold is of the order of four to five times its basal level. Extrapolation of the heat loss curve in Fig. 1.1 (which is roughly representative of an unclad adult human) shows that the heat loss in air below 0 °C would exceed the maximum heat production; and yet, humans and other mammalian species thrive at air temperatures far below the freezing point of water, and normally do not waste fuel on producing extra heat just to keep warm. Insulating the body is the obvious solution, be it by fur in terrestrial animals or by blubber in aquatic mammals. The development of effective **insulation** was the third prerequisite to the evolution of homeothermy. However, insulation becomes increasingly inefficient with decreasing body mass. The consequence is that very small inhabitants of severely cold climates must limit the time of exposure and retreat temporarily to the microclimate of burrows.

A corresponding line of arguments applies to the hot regions of the earth, in which the potential rate of radiant heat flow from the sun into the animal by far exceeds the sustainable rate of evaporative heat loss. Small animals must be nocturnal and spend the hot hours of the day in **burrows**. Larger animals use a less apparent property of **fur**, which offers good protection not just against external cold, but also against radiant heat. It follows that a distinction has to be made between different strategies to cope with thermal stresses. Long–lasting heat and cold require avoidance in small animals and appropriate insulation in larger ones. The effects of **short–lasting thermal stresses** on body temperature, however, can be limited by active adjustments of heat production and heat loss. This is the domain of the **central nervous regulation of body temperature**.

1.3
Homeothermy in Terms of Control Theory

For a first approach, the temperature–regulating circuit can be analyzed by labelling its components with terms borrowed from control theory (Fig. 1.3). The regulated variable of the system is body temperature. It is transduced by temperature sensors in neuronal signals which are fed in a controller and compared to a sort of reference signal or set–point. A difference between the regulated variable and the set–point constitutes a load error and generates efferent drives to appropriate effector mechanisms. The effector responses act in a direction opposite to the load error and tend to limit it: the system is characterized by **negative feedback**.

What body temperature represents the **regulated variable**? In most conditions, a rather stable internal or core temperature contrasts with fluctuating skin temperature. This is often thought to indicate that the temperature of the body core is the regulated variable. However, the most stable temperature in a temperature–controlled system is not necessarily the controlled one. An alternative view is that the integrated temperature of all body regions containing temperature sensors represents the regulated variable. Indeed, **multiple temperature sensors** are present in the body core and in the skin (as part of the shell), and form multiple feedback loops.

Similarly, the system is characterized by **multiple effector mechanisms**. If a primary imbalance between heat production and heat loss has displaced body temperature as the controlled variable, the equilibrium can be restored by variations of **skin blood flow** or evaporation of water (**sweating** or **panting**), both altering the heat loss to the environment. **Shivering** or **non–shivering thermogenesis** increase heat production, and by appropriate **behaviour**, production and loss of heat can be modified to a considerable extent.

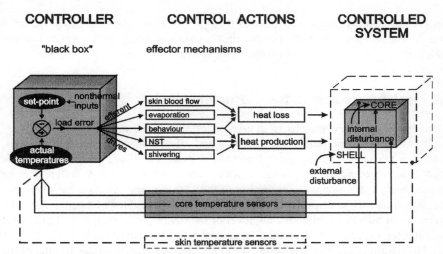

Fig. 1.3. Components of a thermoregulatory system with negative feedback and set–point. The temperature of the body core can be displaced by internal disturbances (such as the waste heat of exercise), while external disturbances (such as exposure to a cold environment) act primarily on the temperature of the skin or shell. In any case, the equilibrium between heat production and heat loss is initially disturbed. The resultant difference between actual temperature and set–point generates efferent drives to effector mechanisms restoring the equilibrium so that deviations of core temperature are limited. (*NST*, Non–Shivering Thermogenesis)

Any regulatory change in heat production or heat loss is generated by and proportional to the load error. An important consequence of **proportional control** is that a constant body temperature under a sustained thermal load requires a sustained difference between regulated variable and set–point. If, for example, a larger heat production during exercise requires a larger heat loss, it can only be initiated and maintained by body temperature exceeding the set–point. Thus, the regulating system can restore the disturbed equilibrium between heat production and heat loss at a higher level of body temperature, but cannot, under a sustained internal heat load, return body temperature to the set–point. The situation is more complex in regard to external cold loads (Chap. 14). Most components and mechanisms of short–term regulation can be interpreted within this frame. Thermoreceptors are the biological equivalents of temperature sensors. Skin blood vessels, sweat glands, skeletal muscle and others serve as effector organs, and the controller function may be intuitively assigned to the central nervous system.

However, the limits of the analogy between biological temperature regulation and the cybernetic control circuit become apparent if one asks for the internal structure of the **central nervous controller**. Theoretically, its functions are to process the integrated signal representing the regulated variable (signal mixer), to compare it to the set–point (comparator), and to generate efferent drives to effector systems, if a difference occurs between regulated variable and set–point. For such detailed analysis, present insights do not provide more than preliminary answers. Consequently, the controller of the type presented above is often dubbed a black box, and the cybernetic control circuit serves primarily as a guide to analyze the mechanisms available to the regulating system.

1.4
A First Approach to the Controller

The function of the controller is, in its broadest sense, to establish the link between afferent signals from temperature sensors and efferent drives to effector systems, and it is performed by the central nervous system. The first attempts to attribute this function to **specific structures within the central nervous system** can be traced back to transection experiments of the 19th century. At least one result has remained largely undisputed by the more refined techniques of later times: total removal of all structures above the level of the rostral brain stem has, if any, very small effects on autonomic regulation against heat and cold [235,282].

The rostral brain stem, however, is of eminent importance. Major **lesions** of the preoptic area–anterior hypothalamus **(POAH)**, or transections below this level, cause permanent functional disturbances. **Single neuron recordings** show that afferent temperature signals from the skin and other extra–cerebral parts of the body core converge on POAH neurons (Chap. 4), making them possible candidates for the integrative function of the controller. **Electrical stimulation** of the posterior hypothalamus generates shivering (Chap. 5), defining this region as the most rostral source of the efferent drive to a specific thermoregulatory effector system. All three arguments show that **neurons of the rostral brain stem** are at least important components of the controller.

Elimination of the POAH, however, does not convert a homeotherm into a poikilotherm. After the POAH was destroyed by proton irradiation, goats still responded to external heat stress with panting [8]. Core temperature had to rise to higher levels than in intact animals, which documents the importance of the POAH for efficient regulation. However, the argument can be turned around: in spite of the POAH being destroyed, the animals were still able to initiate an appropriate, if blunted, response to heat stress. This implies that **other regions** of the central nervous axis **below the level of the PAOH** were also capable of executing the controller function. Supporting evidence comes from experiments in which the **spinal cord** was transected. Local cooling of the spinal cord below cervical or thoracic levels of transection generated shivering in lumbar muscles and reduced skin blood flow in the hindpaws. Both cold defence responses were definitely weaker than in intact animals. The important point, however, is that they could be induced at all. Apparently, the spinal cord not only contains temperature sensors, but all control elements necessary to convey activating signals to some thermoregulatory effector systems. Thus the black box performing controller functions may indeed include the entire neural axis from the lower end of the spinal cord to the rostral brain stem, its different levels perhaps arranged in hierarchical order (Chap.13).

2 The Skin as a Source of Temperature Signals

Experiments have confirmed what an attentive observer may expect: all thermo-regulatory effector mechanisms can be activated or inhibited by changes in skin temperature (T_{skin}). In fact, in many real–life situations T_{skin} can be safely assumed to generate the decisive afferent input. This ensues from its exclusive role in **temperature sensation** [356,428] which is, in contrast to thermal comfort, independent of core temperature (T_{core}). Thus, in a natural environment, changes in T_{skin} often guide thermal behaviour and motivate a subject to seek shelter or adapt clothing so that deviations of T_{core} are reduced or even avoided [212].

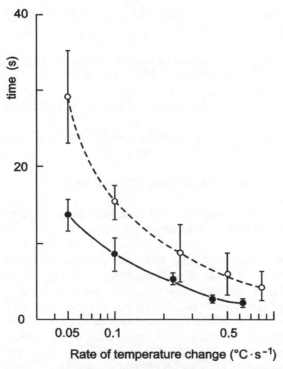

Fig. 2.1. Time taken for human subjects to detect a change in face skin temperature plotted vs. rate of change, during cooling (*solid line*) and warming (*interrupted line*). Symbols and bars means ± SEM. (After [107], with permission)

Skin temperature sensations depend not only on the absolute level of T_{skin}, but also on the degree and **rate of change** from a previous to the actual temperature. Other factors are size and location of the skin region [211]. Figure 2.1 shows the sensitivity to the rate of change: it took human subjects less than 15 s to detect that a 9–cm² area at the cheek had become cooler, in spite of T_{skin} changing as slowly as –0.05 °C·s⁻¹. At higher rates, the time needed for detection decreased considerably. Responses to warming were slower – a fact possibly related to the relatively smaller number of warm sensitive sites of the skin (Chap. 4). The importance of **local T_{skin}** was demonstrated in animal experiments: in a rapidly cycling environment, T_{skin} at the feathered or furred bodies remained unchanged, and the animals' behaviour was entirely determined by the rate of change in T_{skin} at the naked parts [437,438].

With regard to **autonomic effector mechanisms**, the effect of T_{skin} as a source of input signals can be evaluated by relating the **steady–state** activity of cold or heat defence mechanisms to different levels of **constant mean T_{skin}**. The usual way to quantify the responses of effector mechanisms such as heat production, sweating or respiratory heat loss is given by Eq. 2.1:

$$R_{(x)} - R_{0(x)} = \alpha_{(x)} \cdot [T_{skin} - T_{skin}\,thresh_{(x)}] \qquad (2.1)$$

where:

$R_{(x)}$ = response of effector mechanism (x) per unit of body mass or surface area [W·kg⁻¹ or W·m⁻²]

$R_{0(x)}$ = inactive level of effector mechanism (x)

$\alpha_{(x)}$ = proportionality constant or **gain** of effector mechanism (x) [W·kg⁻¹·°C⁻¹ or W·m⁻²·°C⁻¹]

T_{skin} = actual mean skin temperature [°C]

$T_{skin}\,thresh_{(x)}$ = **threshold** of mean skin temperature for effector mechanism (x) [°C]

The equation implies that T_{skin} must pass an **effector–specific threshold** in order to activate or inactivate effector mechanisms: if T_{skin} of a sweating animal is slowly decreasing, sweating ceases finally at the sweating threshold. With further decreasing T_{skin}, another threshold is reached and the animal starts shivering. In an individual, the threshold temperatures depend on a number of factors prevailing in a particular situation which may shift the thresholds a few degrees in one direction or the other. The most important factor is T_{core}: the higher it is, the lower is the threshold of T_{skin} for all effector mechanisms.

If various species are compared, however, it turns out that the thresholds occupy widely different ranges which are mainly determined by the degree of external insulation. Fully fleeced sheep responded with marked **shivering** to a small decrease in T_{skin} at the trunk from 37.4 to 36.2 °C which was caused by transfer from neutral to –9 °C air temperature [522]. In contrast, T_{skin} at the body of bare–skinned swine in Alaska was 11 °C at 0 °C air temperature. However, the animals did not shiver and "made the usual demonstrations of swinish comfort" [248].

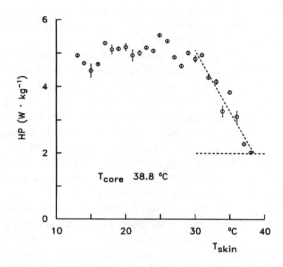

Fig. 2.2. Heat production (*HP*) as a function of skin temperature (*T_{skin}*) at constant core temperature (*T_{core}*). T_{core} of a conscious goat was clamped at the normal value of 38.8 °C. Different levels of T_{skin} were imposed by a water bath in which the animal was immersed to the neck. The *symbols* show means ± SEM of HP in the second half of a 20 min test period. The *horizontal line* shows normal resting HP, corresponding to R0 in Eq. 2.1. The *slope of the second line* represents the gain α of Eq. 2.1, and the *intersection of both lines* determines the threshold temperature. (After [296])

Once the threshold temperature is passed, the autonomic steady–state responses are **proportional to the difference** between actual T_{skin} and threshold. A 4–7 °C decrease in T_{skin} at constant T_{core} usually doubled, by shivering, the heat production (HP) in resting humans [29,42], dog [175,182] and rabbit [150]. The data were estimated on the basis of area–weighted means of locally different T_{skin}, and agree with the results of a study in which shorn goats were immersed in a stirred water bath so that T_{skin} of the entire body except the head equalled water temperature. In the proportional zone, the **gain** was nearly –0.3 $W \cdot kg^{-1} \cdot °C^{-1}$. The resting HP of goat in a thermoneutral environment is approximately 2 $W \cdot kg^{-1}$; HP was therefore approximately doubled by a 6 °C decrease in T_{skin}. At T_{skin} below 25 °C, HP plateaud or even tended to become smaller, implying that the formalization of Eq. 2.1 is restricted to the medium range of effector activity: outside this range, the system is characterized by non–linearities (Fig. 2.2).

In **sweating** humans, a rise in T_{skin} as small as 2 °C above the threshold matched evaporative heat loss with resting HP [42,358], and in all species also **skin blood flow**, the third major autonomic effector system, responds to changes in T_{skin} [171,537]. In **panting species**, however, the response per °C change in T_{skin} was only a small fraction of the capacity of the respiratory heat loss mechanism [150,366]. Because T_{skin} of furred animals changes only to a minor degree upon transition from temperate to warm environments, the importance of T_{skin} in the control of panting tends to be relatively small, except in more sedentary species exposed to intense solar radiation [118].

The question is whether all **skin regions** are of equal thermosensitivity and importance. Studies in which temperature stimuli to different skin areas of normalized size were related to the sweat rate at a thigh showed that this is not the case: the effect of the **face** was three times larger than could be accounted for by its fractional area alone [99,362]. Another somewhat delicate skin region is highly sensitive to heat: warming the **scrotum** of ram to above 36 °C induced panting and a subsequent 1 °C fall in T_{core} [513]. Thus, depending on local properties and **intensity of thermal stimuli**, the temperature of even a small skin region can provide a highly significant component of the total skin signal and, hence, in the regulated variable of the temperature–regulating system.

Sudden exposure to cold, for example entrance into cold water, is often followed by a burst of HP, occurring at constant T_{core} and fading away at the new low T_{skin}. In ramp experiments, changes as slow as -0.1 °C·min^{-1} were observed to induce **dynamic responses to the rate of change** in T_{skin} [296,505]. Such slow changes could conceivably occur also in furred animals in a natural environment. The significance of dynamic sensitivity lies in the fact that it temporarily amplifies the proportional response to increasing thermal stress, anticipating a possible future change in T_{skin} [67]. Related observations were made on sweating in humans. The sweat rate showed a dynamic inhibition if the rate of change in T_{skin} was at least -0.2 °C·min^{-1} [358].

3 The Inner Body as a Source of Temperature Signals

Even small displacements of body core temperature (T_{core}) from its normal level induce large effector responses. Figure 3.1 shows data of experiments in a goat. To dissociate skin temperature (T_{skin}) from T_{core}, the animal was immersed to the neck in a rapidly circulating water bath which clamped T_{skin} at water temperature. Heat exchangers in an arteriovenous shunt were used to manipulate the temperature of the blood, and indirectly the temperature of all core organs and tissues except the skin. Thus, Fig. 3.1 presents responses to pure changes in T_{core} at constant T_{skin}.

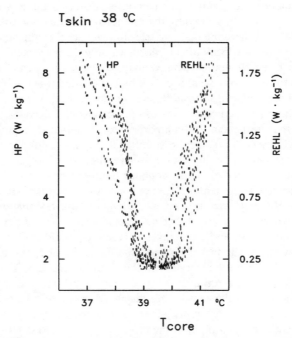

Fig. 3.1. Relationships between core temperature (*T_{core}*), heat production (*HP*), and respiratory heat loss (*REHL*) in a conscious goat. Skin temperature (*T_{skin}*) was clamped at 38 °C by a water bath, while T_{core} was manipulated by heat exchangers acting on blood temperature. A *single symbol* shows 1–min data. (After [366])

Figure 3.1 shows data from four experiments at 38 °C T_{skin}. When T_{core} was lowered from 39.5 to 37 °C, heat production (HP) rose by shivering from its resting level below 2 to nearly 8 $W \cdot kg^{-1}$. When T_{core} was raised to 41.5 °C, respiratory evaporative heat loss (REHL) by panting approached nearly 2 $W \cdot kg^{-1}$. Thus, displacements of T_{core} of the order of ±2 °C around the normal level (the central gap between data points in Fig. 3.1) were sufficient for the full activation of heat or cold defence mechanisms. Similar responses were obtained in other species [27,189] and autonomic effector mechanisms such as sweating and skin blood flow [104,378]. Again as a rule, an increase in T_{core} of the order of 2–3 °C is sufficient to mobilize the full capacity. The magnitude of the responses is entirely proportional to the deviation of T_{core} from its threshold level. So far, attempts to demonstrate dynamic effector responses to the rate of change in T_{core} yielded clearly negative [421] or questionably positive results [344]. Displacements of T_{core} have a major impact also on thermal comfort [43,356].

3.1
Identifying Single Feedback Loops

The question is, **where in the body core** the thermosensitive sites are situated whose signals induce and maintain the responses. Before dealing with present answers, it appears necessary to review the methods by which putative temperature sensors can be traced. Imagine that one tries to evaluate the contribution of a certain site to the overall feedback loop of the thermoregulatory system.

The site may be small, and the sensors may be densely packed. Then a successful approach is to selectively heat or cool it by a so–called **thermode**: a number of water–perfused tubes placed to straddle the site under investigation. For example, heating a sensitive site activates heat loss responses, and because the power required to heat the site is small in comparison to the induced responses, the temperature of the body core outside the stimulated area must fall. On the one hand, this is a safe criterion for having hit a sensitive site. On the other hand, the body core outside the stimulated area also contains temperature sensors, and its secondary temperature deviation in a direction opposite to the primary stimulus must inevitably reduce the response, leading to systematic underestimation [67]. Other problems concern the more or less accurate placement of the thermode around or within the sensitive site, and the uneven temperature field created by the thermode. The consequence is that the determination of the effective temperature of the sensors, which is required to establish the stimulus–response relationship of a site, is subject to some uncertainty.

The problems grow if the sensors are dispersed in a larger mass of unspecific tissue. Then additional power is necessary to heat the entire mass, and the ratio between input and induced output of the effectors is bound to deteriorate. The unevenness of the temperature field within the thermode increases, and the temperature of the sensors remains unknown. It follows that the thermode technique works best if it is applied to a site containing a large fraction of sensors in a small and easily accessible volume of tissue. With decreasing accessibility or spatial density of sensors, the technique runs into problems, and reliable stimulus–response relationships become increasingly difficult to determine.

3.2
Hypothalamus

Since Barbour's study in 1912 [19] on the hypothalamus, no other site has been found in the body core of mammals where temperature changes confined to a small volume of tissue yield such large responses of all thermoregulatory effector mechanisms, including behaviour. Figure 3.2 is from a study by Hellstroem and Hammel [208], whose techniques have set the standard.

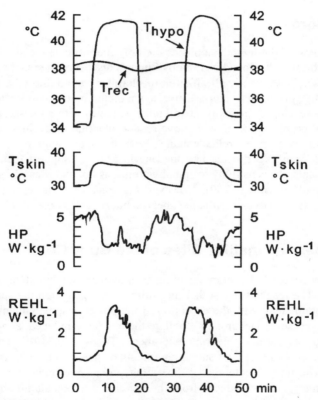

Fig. 3.2. Relationship between hypothalamic temperature (T_{hypo}) and ear skin temperature (T_{skin}), heat production (HP), and respiratory evaporative heat loss ($REHL$) in a dog. T_{hypo} was circled between 34 and 42 °C by a locally implanted thermode. Low T_{hypo} increased HP by shivering and lowered T_{skin} by vasoconstriction, while high T_{hypo} increased REHL by panting, and T_{skin} by vasodilation. T_{core} outside the hypothalamus (T_{rec} rectal temperature) changed in directions opposite to T_{hypo}. Air temperature 25 °C. (After [208], with permission)

Similar observations were made in all mammalian species investigated so far. Both warm and cold stimuli are effective, with better–developed warm sensitivity in larger species. A good criterion is the change in extrahypothalamic T_{core} in a direction opposite to the hypothalamic stimulus because it reflects the combined result of all effector responses: minute displacements of T_{hypo} were followed by opposite changes in general T_{core} [78,191,481]. The sensitive region is small and restricted to the preoptic area and anterior hypothalamus (POAH). Local cooling of the posterior hypothalamus inhibited rather than activated shivering [397].

Several attempts were made to determine the fraction of the total T_{core} input which is generated by hypothalamic temperature sensors. A crude estimate is that, in the standard mammal and in the normal temperature range, **T_{hypo} provides about one half of the total** [259,466]. However, there may be exceptions [200].

3.3
Spinal Cord

The older view of the hypothalamus as the only thermosensitive site of the body core was definitely disproved in 1964 by Simon's discovery of spinal cord thermo-sensitivity [467]. As a rule, **all effector mechanisms available to a species are influenced by spinal cord temperature**, again including behaviour. However, the temperature–sensing elements lie at still unknown sites within the spinal cord, and the technical problems mentioned above render quantitative evaluations of spinal thermosensitivity even less reliable than those of the hypothalamus.

Estimates from experiments like the one in Fig. 3.3 suggest that, at least in larger ruminants, the sensitivity to spinal warming is not significantly smaller than to hypothalamic warming [259]. However, the spinal cord is less responsive to cooling. This is most evident in larger species, but was also observed in rat [18].

3.4
Other Thermosensitive Sites of the Body Core

Thermodes were used to explore the brain stem between the hypothalamus and the spinal cord. The results suggest that the entire neural axis from the spinal cord to the rostral brain stem, with the exception of the posterior hypothalamus, contains temperature sensors. Heating or cooling the medulla induced appropriate responses of behaviour and shivering, and altered T_{core} [309,310]. Thermal stimulation of the midbrain caused subtle alterations of heat production and heat loss [100,447]. The limited potential of the extrahypothalamic brain stem, at least in larger mammals, can be deduced from the fact that a hypothalamic thermode detected 80% of the temperature sensitivity of the goat's entire brain [191,298].

Core temperature sensors are not restricted to the central nervous system. Local warming of the intestinal walls in sheep [401] or the dorsal walls of the abdomen in rabbit [410] modulated HP, panting and skin blood flow. However, the responses did not indicate a high spatial concentration of temperature sensing elements at both sites. This was also the case for skeletal muscle, local cooling of which inhibited panting [264].

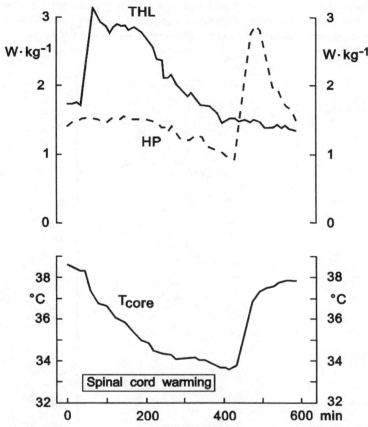

Fig. 3.3. Temperature sensitivity of the spinal cord. At an air temperature of 15 °C, the spinal cord of an ox was selectively warmed by a chronically implanted peridural thermode, while total heat loss (*THL*) and heat production (*HP*) were measured by direct and indirect calorimetry. Spinal warming induced a long–lasting increase in THL by sweating, panting and high skin blood flow, causing T_{core} (taken at the tympanic membrane) outside the spinal cord to decrease. (After [268], with permission)

Data obtained with thermode techniques show generally a mismatch between the rather small responses to cooling single sites, and the strong effect of cooling the entire body core. This points to the presence of temperature sensors in other regions. One possibility is that **sensors are dispersed throughout the trunk**, which would preclude tracing by thermodes in conscious animals.

An alternative is to clamp T_{brain} and $T_{spinal\ cord}$ at constant levels, and to change the temperature of the residual body by manipulating arterial blood temperature. This method affects all temperature sensors of the body, except those in the clamped regions (Fig. 3.4). Results of such studies support the hypothesis that a substantial part of the thermal responses to low T_{core} is generated by hitherto undetected sensors located outside the central nervous system [265,335].

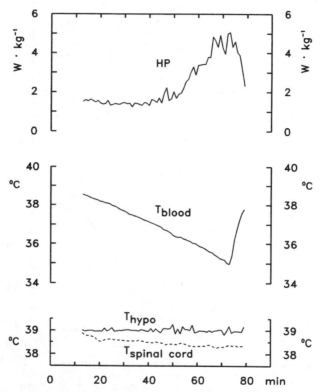

Fig. 3.4. Effect of cooling the residual body core at constant brain and spinal cord temperature. In a goat, the temperatures of brain (*T_{hypo}*) and spinal cord (*T_{spinal cord}*) were clamped between 38 and 39 °C, while the arterial blood supplying the trunk (*T_{blood}*) was cooled. With decreasing temperature of the residual body core, shivering occurred and heat production (*HP*) increased. Air temperature 20 °C. (After [265])

4 The Neuronal Basis of Temperature Reception

4.1
Skin Thermoreceptors

When skin of a human subject is systematically explored with a pointed thermode of 28 °C or lower, certain small sites yield clear cold sensations while others do not. Similarly, if the thermode temperature is 40 °C, some sites generate warm sensations. Cold spots are more frequent than warm spots, and the density of both types varies, being high in the face and low in the calf [211].

In a similar experiment, the discharge rate of a specific afferent neuron from the skin or the tongue changes when a thermode alters the temperature of its spot–like receptive field. Two receptor types can be distinguished, consistent with the two qualities of sensation: **cold and warm receptors**. The clearest difference is in the **dynamic properties** (Fig. 4.1). Cold receptors respond to a sudden drop in skin temperature (T_{skin}) with a short–lasting increase in activity, but may become even silent for a few seconds when T_{skin} is rising steeply. Conversely, warm receptors overshoot during warming and show transient inhibition during rapid cooling. The dynamic responses are far from being small: when T_{skin} was quickly raised from 38 to 42 °C, the discharge rate of a warm receptor jumped briefly from $9 \cdot s^{-1}$ to $50 \cdot s^{-1}$ before settling at $20 \cdot s^{-1}$ [215]. Also the **static discharge rates** of cold and warm receptors at constant T_{skin} depend on its level.

The **morphology of cold receptors** was described in detail. In the hairy skin of the cat's nose, the receptive structures at cold spots were served with thin myelinated axons, dividing into several unmyelinated terminals. The terminals contained numerous mitochondria and penetrated a few μm into the basal layer of epidermal cells [213]. Most cold receptors were found approximately 0.2 mm below the skin surface. However, there is good evidence, at least in rabbit, that specific cold receptors occur also at deeper levels, near the subcutaneous fat layer [253].

Neurophysiologically, each of the two types of receptors has almost uniform properties among different skin areas in the same species and among species, with few exceptions [202]. The conduction velocity of cold receptor afferents is in the range from 5 to 10 $m \cdot s^{-1}$, and that of the nearly unmyelinated warm receptor afferents around 1 $m \cdot s^{-1}$ [211]. In most areas, cold receptors are much more frequent than warm receptors. The exception is the scrotum [243].

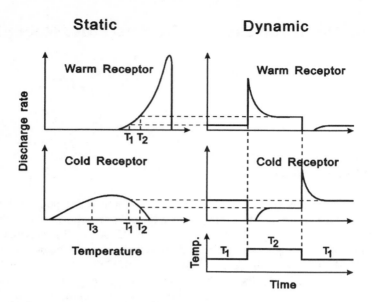

Fig. 4.1. Properties of skin thermoreceptors. *Right* A sudden increase in T_{skin} from T_1 to T_2 induces, for a few seconds, an overshooting discharge rate of a warm receptor, while a cold receptor is completely inhibited. The pattern is reversed when T_{skin} decreases from T_2 to T_1. *Left* At constant temperatures, both receptor types show constant discharge rates, with roughly bell—shaped relationships between temperature and discharge rates. The static response is ambiguous: T_1 and T_3 result in the same discharge rate of a cold receptor. (After [214])

The **mechanisms of temperature transduction**, through which the receptors generate temperature–dependent static discharge rates and dynamic responses, are still enigmatic. One reason is that the receptive terminals are too small to be explored with current methods of membrane physiology. As a substitute, temperature–dependent membrane properties of large invertebrate neurons were studied. In pacemaker neurons of *Aplysia*, the electrogenic sodium pump and the passive sodium/potassium permeability ratio are highly dependent on temperature. Both properties are common to all excitable structures and could modulate an intrinsic oscillation frequency of a receptor potential [58,85]. There is good evidence that at least the temperature–dependent electrogenic sodium pump plays an important role also in mammalian thermoreceptors [392].

The **bell shape of the static curves** poses another unsolved problem. The mean discharge rate of a population of cold receptors attains its maximum near 29 °C, and declines with increasing and decreasing temperature. One consequence is that different skin temperatures, for example 20 and 35 °C, result in the same mean discharge rate of the primary afferents (Fig. 4.1). This is difficult to reconcile with common experience: both temperatures yield clearly different sensations. Also cold defence responses like shivering do not become smaller, when T_{skin} decreases below 29 °C (Fig. 2.2). Thus, the mean discharge rate of a population of

primary cold afferents does not reflect the skin–temperature signal arriving at the
sensory and integrative levels of the central nervous system. However, a larger
population of cold receptors is composed of units having widely different individual temperature–frequency curves, with maxima at temperatures between 10 and
40 °C. The **across–fibre pattern** of the population rather than the mean discharge
rate could encode unequivocal information over the full range of T_{skin} encountered under natural conditions [211].

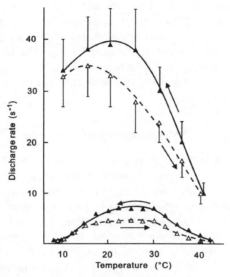

Fig. 4.2. Thermal afferents from the face. Mean responses from 18 trigeminal ganglion recordings (1st–order neurons *bottom*) and 18 trigeminal nucleus recordings (2nd–order neurons *top*)
plotted versus the seemingly constant temperature of the receptive fields in rat. *Closed symbols*
show data from cooling sequences, and *open symbols* from warming sequences. A *single symbol*
shows mean and SEM of six 2.5 s epochs, after the discharge rates were judged constant. SEM
on *bottom curves* were smaller than symbols. (After [108], with permission)

As mentioned above, specific cold and warm receptors are highly sensitive to the
rate of change in temperature. These dynamic properties underlie the rapid detection of even small changes in T_{skin} (Fig. 2.1), and are instrumental in guiding
thermal behaviour. Of relevance to autonomic temperature regulation (Chap. 2) is
the question, whether rather **slow changes in temperature** still induce dynamic
receptor responses. The answer was positive for rates as low as –2.4 °C·min^{-1}
[346], and the results of Fig. 4.2 imply a dynamic sensitivity to even slower
changes. The figure shows, for first–order cold afferents in the trigeminal ganglion
and second–order afferents in the trigeminal nucleus, a large hysteresis. Particularly in the medium range, the discharge rates at a certain, seemingly constant
T_{skin} were high during a cooling sequence, and low during a warming sequence.

This occurred in spite of the rates having been judged constant for 60 s before the data were sampled. Similar observations had been reported before, with the annotation that the hysteresis disappeared after the adapting period had been extended to 8–10 min [119].

Temperature signals from the skin are **subject to efferent control**. It is exerted via the sympathetic system and acts apparently at the level of the primary afferents. Electrical stimulation of the cervical sympathetic chain excited cells in the trigeminal ganglion and nucleus which were sensitive to cold stimuli in the face. The excitation was resistant to ß–blockade and could be mimicked by α–agonists [106]. Thus, higher sympathetic activity could sensitize cold receptors and generate, at the central integrative level of the thermoregulatory system, a peripheral cold signal larger than warranted by the actual T_{skin}. The biological significance of the effect is not yet clear; it is perhaps related to the observation that general sympathetic activation can be associated with elevated internal body temperature (emotional fever or stage fright).

4.1.1
Afferent Processing of Skin Temperature Signals

Another aspect of Fig. 4.2 is that, at any given T_{skin}, the discharge rate of second–order neurons in the trigeminal nucleus was about fivefold higher than of first–order neurons in the trigeminal ganglion. Because the second–order receptive fields were also much larger than the spot–like first–order ones, the ascending pathway is characterized by **spatial summation**. No other evidence of signal processing was observed, and most neurons, if not all, projected directly to the thalamus. Similar observations were made in humans [109]. Thus the thalamus receives a basically undistorted copy of the **face skin temperature** signal, as it is encoded in the discharge rate of the primary afferents, and serves as relay site to the cortex. No hard evidence is available as to how and in what form the face skin temperature signal reaches hypothalamic integrative structures.

Peripheral afferents from the skin of the **trunk** and the limbs terminate in the spinal cord, at the superficial laminae of the dorsal horn. From segmental levels, at least one pathway leads via the anterolateral quadrant of the cord to the pontine **nucleus raphe magnus**. The temperature–response curves of the raphe neurons in Fig. 4.3 are similar to those of skin cold and warm receptors, even though the shifted range and higher proportion of warm reactive neurons suggests a certain degree of processing, in addition to spatial summation. However, different levels of internal body temperature had no effect on the relationship between T_{skin} and activity, implying that N. raphe magnus receives and transmits pure skin temperature signals [115]. From N. raphe magnus, the pathway projects to the midbrain raphe nuclei and reticular formation, and diverges further to sensory thalamic nuclei and regulatory hypothalamic areas [202,392].

In the dorsal horn and at higher levels, neurons responding to limb and trunk skin temperature are rare, while neurons responding to the temperature of the **scrotum** are abundant. The exceptional representation of the scrotum may be related to the importance of testes temperature for fecundity and fits in with the large panting response to local heating of the scrotum (Chap. 2). The scrotal afferents take the same routes as the ones from the trunk and the limbs, but are characterized by a high degree of signal processing, which includes massive bilateral convergence [205]. A unique feature is the generation of a **switching pattern** in N. raphe magnus: the static response range was reduced from approximately 10 °C in first–order neurons [203] to 0.5 °C or less in raphe, thalamus and hypothalamus [206]. In other words, the higher–order neurons changed from minimum to maximum activity within a 0.5 °C range of scrotal temperature. Individual neurons switched at different temperatures. However, the percentage of neurons discharging at maximum rate increased with temperature so that scrotal temperature appeared to be represented centrally by the fraction of switched neurons [204].

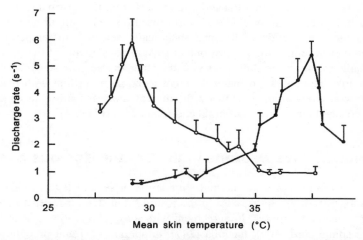

Fig. 4.3. Representation of skin temperature in the pontine N. raphe magnus. Mean activities of 17 cold–reactive neurons (*open circles*), and 25 warm–reactive neurons (*filled circles*) plotted vs. mean skin temperature in rats. *Bars* SEM. (After [115], with permission)

4.2
Core Temperature Sensors Outside the Central Nervous System

These elements share a property with skin thermoreceptors: the process of temperature transduction occurs at the peripheral endings of neurons whose cell bodies are located in spinal dorsal root ganglia or ganglia of cranial nerves. Studies dealing with putative peripheral sensors generally encounter problems in assigning a specific thermosensitive function to a neuron [369]. A variety of mechanical and chemical stimuli has to be ruled out before a neuron can be characterized as a specific thermoreceptor. However, it is still an open question whether strict unimodality is an indispensable prerequisite for neurons to be involved in the regulation or sensation of temperature [285]. Thus, the observations quoted below may indeed show the neuronal basis of the afferent generation of regulatory responses which can be induced by local or general temperature changes of the body core.

Putative temperature sensors within the body core but outside the central nervous system were first found in tissues enclosing the suprarenal gland and the root of the superior mesenteric artery of rabbit. The afferent fibres were part of the splanchnic nerve and displayed static and dynamic responses comparable to skin warm receptors [408]. Similar recordings were obtained from the hepatic branch of the vagal nerve in guinea pig. These units were specifically responsive to the temperature of the liver [1]. In cat, cold fibres in the splanchnic nerve responded to cooling of the stomach and the adjacent duodenum [160], and afferents in the pelvic nerve were excited by local cooling of the bladder [307]. In the same species, C–fibres in dorsal roots transmitting signals from the gastrocnemius muscle behaved like afferents from cold and warm receptors [332]. In all nerves investigated so far, the fraction of neurons specifically sensitive to temperature was small. However, temperature is just one of several constituents of the internal milieu signalled to the central nervous system. Thus, the very presence of temperature–sensitive neurons at so many different locations is in line with results obtained in conscious animals showing thermoregulatory responses to cooling the body at constant temperatures of brain and spinal cord (Fig. 3.4).

4.3
Thermosensitive Neurons in the Central Nervous System

Chapter 3 showed that local thermal stimulation at several levels of the central nervous axis, from the spinal cord to the rostral brain stem, can generate specific thermoregulatory responses. In agreement with these observations, neurons were found in spinal cord [464], medulla [247], midbrain [101] and preoptic region–anterior hypothalamus (POAH), whose discharge rates varied more with the **local temperature** than those of other neurons [53]. These **thermosensitive neurons** could present the central nervous equivalent of skin thermoreceptors. However, at present no neurophysiological or other evidence exists of primary thermoreceptors in the central nervous system, that is, of units having no afferent input. An alternative is that central **neurons in the chain** from peripheral thermoreceptors to efferent pathways possess a particular degree of thermosensitivity. Most studies used extracellular recordings in the POAH of anaesthetized animals. In larger samples, about 30% of the neurons responded to local warming with increasing discharge rates (warm–sensitive). The discharge rates increased in 10% during local cooling (cold–sensitive), and about 60% were considered insensitive because the temperature dependence of discharge rates did not surpass **arbitrarily defined limits** [52]. A fraction of central neurons also responded to the **temperature of remote regions** of the body such as the skin, and are termed **thermoreactive** [503]. This implies that a central neuron can be thermosensitive and thermoreactive; potentially, these neurons could serve simultaneously as local temperature **sensors and integrators** of local and peripheral thermal information [52].

4.3.1
Neuronal Temperature Transduction

More recently, a new in vitro technique was introduced. Thin slice preparations of hypothalamic tissue, surviving for some time in a nutrient medium, permit intracellular recordings and provide a tool to study temperature effects on neuronal membranes.

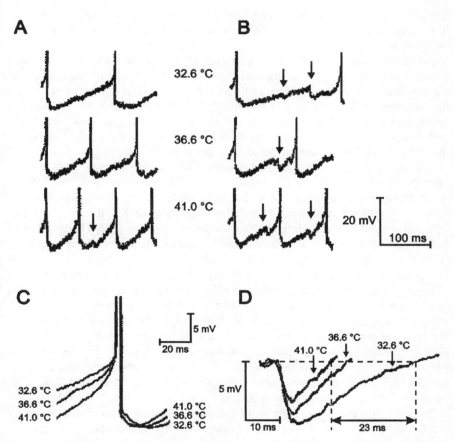

Fig. 4.4 A-D. Temperature–dependent pacemaker potentials of POAH neurons in rat. In tissue slices, all spontaneously active neurons display depolarizing prepotentials which generate action potentials. In some neurons, the slope of the prepotentials depends clearly on temperature (*single recordings* in **A** and **B**, action potentials truncated). **C** Each trace shows the average of 15 prepotentials superimposed on threshold. **D** Averaged IPSPs (*arrows* in **A** and **B**), showing that warming decreases amplitude and duration of postsynaptic potentials. (After [52], with permission)

Figure 4.4 shows the sensitivity of a slice neuron to warming of the medium: the higher the temperature, the shorter was the interspike interval (Fig. 4A, B). Between the spikes, depolarizing prepotentials, or **pacemaker potentials**, rose slowly to the threshold at low temperature, but faster at higher temperature so that it required less time to reach the threshold. The inherent thermosensitivity of the neuron itself was supported by inhibitory postsynaptic potentials (**IPSPs**) from nearby **temperature–insensitive neurons** which, in vivo, could be conceived as transmitting thermal information from the skin and other remote sites of the body. Amplitude and duration of the IPSPs were reduced at higher temperature so that their effectiveness to delay the prepotentials decreased with increasing temperature. Thus, both the temperature dependence of their own prepotentials and of the IPSPs from nearby thermoinsensitive neurons constituted the greater thermosensitivity of some neurons [52].

An important point is that the prepotentials of all spontaneously active neurons in the slices depended on temperature: the difference between insensitive and warm sensitive is quantitative rather than qualitative, and no self–evident borderline between categories could be established [157]. A similar degree of intrinsic thermosensitivity, as judged from the temperature dependence of discharge rates, was displayed by cells in slice preparations containing the superficial laminae of the spinal dorsal horn, where afferents from the skin terminate [390].

The question is of course whether neurons like the one of Fig. 4.4 belong specifically to the temperature–regulating circuit. The **temperature sensitivity of discharge rate** is no proof: in the sensorimotor cortex of cat, an area not known to have any function in autonomic temperature regulation, 48% of the neurons (most certainly pyramidal tract cells) were highly thermosensitive [20]. The **location** of a thermosensitive neuron in an area known to induce thermoregulatory responses to local temperature might be a better criterion. However, the sites of the rostral brain stem which generate thermoregulatory responses to local temperature host also non–thermal circuits, and thermosensitive neurons at such sites might be connected and introduce some degree of unspecific temperature sensitivity. Indeed, POAH neurons deprived of all afferent inputs responded not just to temperature but to a variety of other stimuli such as osmolality and glucose concentration [367,459]. Neurons in the central nervous system may generally possess multimodal sensitivity to their local environment – a feature providing a simple explanation for temperature effects on non–thermal control circuits, but a great obstacle to the analysis of the function of a neuron in a single circuit. Thus, even the combination of temperature sensitivity and location of a neuron is no conclusive argument in favour of its involvement as a specific sensor in the regulating circuit.

4.3.2
Sensitivity to Local and Remote Temperatures

There is a third criterion: if a neuron is thermosensitive and thermoreactive, that is, responding to its own temperature and to the temperature of remote regions, then its involvement in the thermoregulatory circuit appears more likely, and it may serve as a local sensor. Such neurons have indeed been found. At least in the spinalized cat, all units sensitive to spinal cord temperature responded also to T_{skin} (Fig. 4.5).

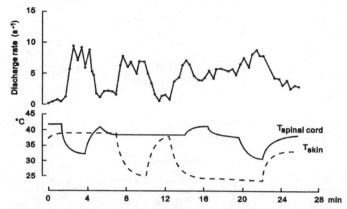

Fig. 4.5. Discharge rate of an ascending spinal neuron in a cat. The neuron was sensitive to the temperature of the spinal cord ($T_{spinal\ cord}$), and reactive to skin temperature (T_{skin}). (After [460])

Another thermosensitive region of the neural axis is the medulla oblongata, whose local thermal stimulation induces appropriate autonomic and behavioural effector responses. The medulla contains a large number of locally thermosensitive neurons, and more than one half were found to react also to changes in T_{skin} [247]. The ascending pathways continue to midbrain raphe nuclei and the reticular formation. There again, local thermal stimulation generates effector responses, and recordings in the midbrain showed neurons responding to local temperature and the temperatures of the skin [368] or the spinal cord [234]. Not surprisingly, some neurons in the POAH showed dual sensitivity to local and skin temperatures [532], and a few neurons even had triple sensitivity to local, spinal cord and skin temperatures [54]. It requires some stretch of imagination to assume that these neurons were not part of the temperature regulating circuit (Fig. 4.6).

However, a caveat comes from the other homeothermic class. The POAH of most birds, in contrast to mammals, possesses no specific thermosensory function in the hypothermic range: local cooling inhibits rather than stimulates cold defence mechanisms [466,469]. Nevertheless, the POAH of duck contains many locally warm–sensitive neurons, whose temperature–frequency relationships were indistinguishable from mammalian POAH neurons [463]. The majority of these neurons reacted also to the temperature of the remote body core [306], suggesting involvement in temperature regulation. Thus, putatively integrative neurons of the thermal POAH network can be locally thermosensitive without having thermosensory functions.

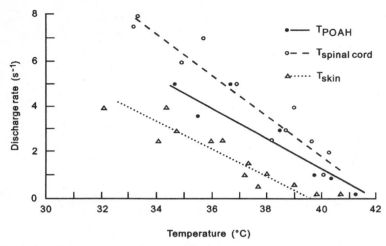

Fig. 4.6. Discharge rate of a neuron in the preoptic area–anterior hypothalamus (*POAH*) of a rabbit. The neuron was sensitive to POAH temperature, and reactive to temperatures of spinal cord (*T_spinal cord*) and skin (*T_skin*). (After [54], with permission)

An interesting observation is that warm sensitive neurons, that is, neurons responding to increasing local temperature with increasing discharge rate, often display the same type of response to remote temperatures. This gave rise to the concept of warm signal vs. cold signal networks, which were supposed to activate the appropriate effector mechanisms [233]. However, the rule of equidirectional responses has numerous exemptions, and others hold that true core temperature sensors are entirely of the warm type, while cold–sensitive neurons are supposedly integrative neurons [51].

In summary, the neurophysiological basis of specific central nervous thermosensitivity is still far from being understood. The main reason is uncertainty about whether a particular neuron in the diffuse system of the rostral brain stem is part of the thermoregulatory circuit, and if so, whether it is on the input, integrative or output side of the network [202].

5 Heat Production and Heat Balance of the Body

5.1 The Heat Balance Equation

As mentioned in Chapter 1, the rates of heat production and heat loss must be equal if body temperature is to remain constant. In this context, the term body temperature refers to the mean of the entire body, and must not be confused with that of the body core. Thus, in its simplest form, the heat balance equation for the case of constant mean body temperature of a resting subject is:

$$HP = THL \tag{5.1}$$

where:

HP = rate of metabolic heat production

THL = rate of total heat loss

Dimensions are watt (W), preferably in relation to unit area of body surface ($W \cdot m^{-2}$) or unit body mass ($W \cdot kg^{-1}$).

A more detailed form of Eq. 5.1 is:

$$S = MR - (\pm W) - (\pm E) - (\pm C) - (\pm K) - (\pm R) \tag{5.2}$$

where:

S = rate of storage of heat (positive = increase in body heat content, negative = decrease in body heat content)

MR = rate of metabolic energy transformation (always positive in a living animal. During rest MR = HP, during positive work MR = HP + W)

W = rate of work (positive = external work accomplished, negative = mechanical work absorbed by the body)

E = rate of evaporative heat transfer (positive = evaporative heat loss, negative = evaporative heat gain)

C = rate of convective heat transfer (positive = transfer to the environment, negative = transfer into the body)

K = rate of conductive heat transfer (positive = transfer to the environment, negative = transfer into the body)

R = rate of radiant heat transfer (positive = transfer to the environment, negative = heat absorption by the body)

All again in $W \cdot m^{-2}$ body surface area or in $W \cdot kg^{-1}$ body mass.

In many situations, all components of heat loss are positive, because the environment is usually cooler than the body. However, they may eventually become negative. The most common case is a net radiant heat gain caused by solar radiation. Also situations involving convective and conductive heat gains are, at least in humans, far from unusual, as visitors of a hot water bath can testify. In contrast, a net evaporative heat gain is a rather exotic experience, requiring condensation of water vapour from the surrounding air on the skin. Finally, external work can involve a negative component when, for example, mechanical energy is absorbed as heat in walking downhill [503].

An important point is the storage or destorage of heat, occurring when all other components of the heat balance equation do not cancel each other. An imbalance of $1 \ W \cdot kg^{-1}$ between HP and THL may persist in a resting subject for 1 h:

$$HP - THL = \pm 1 \ W \cdot h \cdot kg^{-1} = \pm 3.6 \ kJ \cdot kg^{-1} \qquad (5.3)$$

For convenience, it is further assumed that the specific heat of the body (the quantity of heat required to raise the temperature of unit mass by 1 °C) be:

$$c = 3.6 \cdot [kJ \cdot kg^{-1} \cdot °C^{-1}] \qquad (5.4)$$

It follows that a difference of $1 \ W \cdot kg^{-1}$ persisting for 1 h must change mean body temperature by 1 °C up or down. In contrast to the simplification of Eq. 5.4, the specific heat of the body is usually taken to be 3.47 $kJ \cdot kg^{-1} \cdot °C^{-1}$. However, the average specific heat of a body depends on its composition because the specific heat of its components is different: from 2.12 of fat to 4.18 of water. Thus, the true average is between 3.2 for a lean animal containing 12% body fat, and 2.7 for an obese animal containing 50% fat [41].

The precise determination of changes in mean body temperature requires simultaneous measurements of HP by indirect calorimetry, and of THL by direct calorimetry [329]. Provided that environmental conditions and food intake of a resting animal are constant, short–lasting imbalances between the production and loss of heat mainly reflect the circadian rhythm of body temperature and tend to cancel each other over 24–h periods so that the storage term becomes negligible [327].

However, environmental thermal loads can induce significant changes in **mean body temperature**, which are not necessarily reflected in changes in body core temperature. In a warm environment, the **thermal core** of an animal is large and surrounded by a thin shell. In the cold, however, the warm core contracts and becomes increasingly insulated by a thicker **shell**, the outer temperature of which approaches the temperature of the environment. The underlying effector mechanism is a change in peripheral blood flow. This important point is treated in more detail in Chapters 10 and 14. However, in the context of the heat balance equation, it is evident that a **transition** from a warm to a cool environment is usually accompanied by **destorage of heat**. Conversely, a transition to a warm environment involves initial **storage** of a considerable amount of heat in tissues which are part of the shell in the cold, and become part of the core in the warmth.

5.2
Body Mass, Heat Exchange and Metabolic Rate

The body mass of terrestrial placentals ranges from less than 10 g in shrews to 6000 kg and more in elephants. In animals of similar shape, the surface area increases with the 2/3 power of body mass [331]. Thus the **surface area per unit of body mass** is largest in the smallest animal, and decreases with increasing mass (Fig. 5.1, top). Because the heat exchange with the environment is proportional to the size of the surface area, it follows that, provided all other variables are equal, the **rate of heat exchange per unit of body mass** is also largest in the smallest animal. Thus, in a given cold environment, a rat of 250 g body mass loses heat at a much higher rate than a 250–g fraction of an ox. Conversely, small species exposed to intense solar radiation gain heat at higher mass–specific rates than large ones. The same relations hold, of course, for small and large individuals of a species, and yet, the normal unstressed internal body temperature of placentals is independent of body mass, that is, in the range 36–40 °C, if one leaves out the very small and the very large, in which measurements are extremely difficult [352].

The burden to the temperature–regulating system of small mammals in the cold is relieved to some extent by the fact that the basal metabolic rate (BMR), that is, the rate of heat production of a fasting resting animal in a thermoneutral environment depends also on body mass [66,289]. Measurements in many species from mouse to elephant have shown that the **basal metabolic rate per unit of body mass** is inversely related to body mass (Fig. 5.1, middle).

Conventionally, this is expressed in an equation relating the BMR of whole animals to their body mass:

$$BMR = k \cdot m^{3/4} \tag{5.5}$$

where: BMR = basal metabolic rate = basal rate of heat production
                      m = body mass
                      k = 3.4, if m in kg and BMR in W

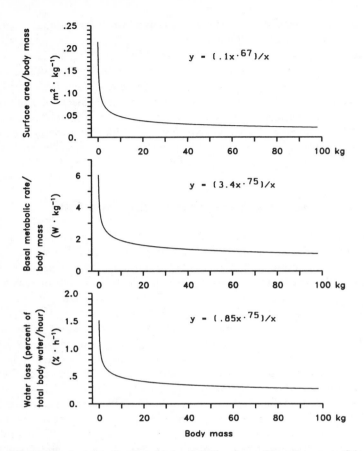

Fig. 5.1. Effects of body mass. *Top* In animals of similar shape, the surface area of the body per unit of mass increases with decreasing mass [331]. *Middle* The basal metabolic rate per unit of mass increases with decreasing mass [289]. *Bottom* The percentage of total body water content, required per hour for complete dissipation of heat generated at basal metabolic rate, increases with decreasing mass

The value of k = 3.4 is a gross average of all mammals [289]. However, considerable differences exist between different taxonomic groups [533]. In marsupials k is about 2.4, resulting in the BMR being 30% lower than in placentals of equal mass, and in a 2–3 °C lower unstressed body temperature. In monotremes, k and BMR are still smaller, and the body temperature is in the range 30–32 °C [112,113]. If the BMR were exactly proportional to the surface area, it would also correspond to the 2/3 power of body mass. However, the 3/4 power is experimentally well established. It is purely empirical and owes nothing to theoretical considerations [41]. The difference shows that factors other than the surface rule additionally influence the relationship between body mass and BMR. However, this does not preclude the ratio of surface area to body mass of a species or individual from being an important determinant of its metabolic rate.

A negative aspect of the relatively larger rate in smaller species is that the **evaporation of water** increasingly loses its expedience as a regulatory means of heat dissipation. Consider an environment in which all avenues of dry heat loss are blocked by high ambient temperature, so that evaporation remains the last resort. In a human being of 80 kg, the evaporation of 140 ml water per hour (approximately 0.3% of total body water content) is sufficient to dissipate the heat generated by the BMR. However, a 100–g animal would have to spend 1.5% of its water content per hour to accomplish the same heat loss (Fig. 5.1, bottom). The situation of small animals on the ground of a hot desert would be even worse. Again owing to the relatively larger surface area, the **radiant heat gain per unit of body mass** is also larger. Returning the gain to the environment by evaporative means would pose another great drain on the body water store, the maintenance of which is the major problem in arid regions.

5.3
Metabolic Heat Production vs. Metabolic Rate

Metabolic heat production (HP) is the rate of transformation of chemical energy into heat. In contrast, metabolic rate (MR) is defined as the rate of transformation of chemical energy into heat and mechanical work. Apparently, HP and MR are equal in a resting animal. During work on an external system, however, MR equals the sum of HP and work [503]. On transition from rest to intense exercise, MR multiplies several times, and such rates can be sustained for considerable periods of time. However, the mechanical efficiency of work, that is, the ratio of work to the exercise–induced increment in MR, rarely exceeds 0.2, while the major part of the increment is released as heat [492]. Its complete transfer to a hot environment may be impossible, in which case a continuous increase in body temperature finally enforces the cessation of exercise. In the cold, however, heat as the by–product of activity decreases the level of ambient temperature, below which shivering or non–shivering thermogenesis must step in to prevent body temperature from falling.

5.4
Basal Heat Production vs. Resting Heat Production

Basal MR or HP is the rate of energy transformation in a rested, awake and **fasted state** in the absence of thermal stress. Its major fraction is produced in the metabolically active organs of the body core, while the contribution of inactive musculature is relatively small. In an individual, the basal HP is tightly coupled with the thyroid status. Triiodothyronine (T3), the biologically active hormone, exerts a variety of stimulating effects on different metabolic steps so that hypothyroidism can result in a 30% reduction in basal HP, while hyperthyroidism is associated with increases of up to 50% [135]. Both changes are unquestionably of thermoregulatory significance. In small species, seasonal changes in tissue levels of T3 are instrumental in adaptation to cold [221].

In animals, the condition of rest is difficult to meet, and the term basal HP is often replaced by minimum HP. Thus, minimum activity substitutes for rest. Furthermore, the definition of resting HP includes the absence of thermal stress, but dispenses with the condition of a fasted state. This is an important point, because the **resting heat production increases with the rate of food intake**. The increment in HP is termed postprandial excess heat production. It is caused by a variety of factors associated with the breakdown of nutrients and synthesis of stored compounds, and particularly large in high–protein diets [41].

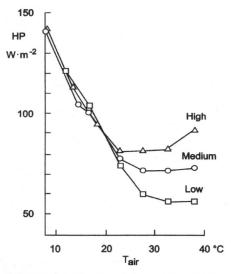

Fig. 5.2. Heat production (*HP*) of a closely clipped sheep at three feeding levels (*Low, Medium, High*) plotted vs. air temperature (*T_air*). *A single data point* shows mean HP over a 72 h period. (After [154], with permission)

From a thermoregulatory point of view, it is the magnitude of the effect that matters most. Figure 5.2 shows the HP of a closely clipped sheep at low, medium and high feeding levels (600, 1200 or 1800 g dried grass cubes·day^{-1}) plotted vs. air temperature. The minimum HP at each feeding level can be taken as resting HP. It was 55 W·m^{-2} at low feeding, and the animal had to actively increase HP already at 28 °C air temperature. At the high feeding level, however, minimum HP was 80 W·m^{-2} and sufficient to balance heat loss even at 23 °C air temperature. Thus, the high food intake during cold exposure diminished the demand for metabolic cold–defence mechanisms. The **lower critical temperature**, that is, the ambient temperature below which HP must increase by shivering or non–shivering thermo-genesis to maintain thermal balance, depends on **food intake**.

It is worth noting that the resting HP can undergo an annual cycle. In some larger species genetically adapted to regular scarcity of food in winter, food intake and, consequently, HP, are smaller in winter than in summer even when food is freely available [131,318,380]. This is clearly counterproductive with regard to metabolic cold defence, and points to external insulation as the primary strategy in these species. In small species, however, insulation is a less feasible means of protection, and resting HP is higher in winter [411].

5.5
Shivering

Shivering is an involuntary tremor of skeletal muscle, and is considered here ex-clusively as a thermoregulatory effector mechanism for increasing the production of heat. The analysis of the tremor is the domain of electromyography: recordings of muscle action potentials from implanted or surface electrodes. The initial re-sponse to slowly increasing cold exposure consists in continuous trains of action potentials from single motor units, which precede overt shivering and correspond with increased muscle tension. The second phase shows **group discharges**: the motor units of a muscle fire in synchrony, and the frequency of the groups equals the frequency of the movements of the muscles or limbs. At the limbs, agonists and antagonists contract in synchrony at low intensity of shivering, but are recip-rocally innervated at high intensity [23]. The **rhythm** is generated in the spinal cord, and its frequency depends on body mass (Fig. 5.3): it is of the order of 10 Hz in adult humans, and 40 Hz in mouse [290].

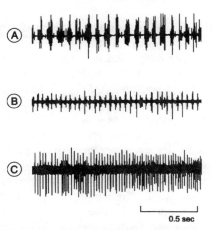

Fig. 5.3 A-C. Electromyograms of grouped discharges in hindleg flexor muscle during mild shivering. The frequency of the discharges depends on body mass and was approximately 12 Hz in dog (**A**), 24 Hz in guinea pig (**B**) and 40 Hz in mouse (**C**). (After [477])

The central efferent drive originates in the dorsomedial region of the posterior hypothalamus [280,504]. It descends, via the shivering pathway, through the reticular formation of the lower brain stem and the lateral white columns of the spinal cord to segmental networks. The efferent drive is a **tonic signal**: the discharge frequency in the descending fibres mirrors the central drive, but has nothing to do with the frequencies of the group discharges and muscle contractions [209].

The maximum shivering HP during cold exposure, in terms of multiples of resting HP, varies between 3 and 10 in different species, including monotremes and marsupials [27,112,411]. As a rule of thumb, maximum shivering is attained when internal body temperature is displaced 2–3 °C below its normal level [189,506], and can be maintained, with little evidence for fatigue, for at least two hours [86]. With further developing hypothermia, HP begins to decline (Fig. 19.2). In this range, the temperature dependence of peak HP in homeotherms resembles that of basal HP in poikilotherms (Chap. 1). It is a good example of Arrhenius–van't Hoff effects in homeotherms which are normally, under less strenuous conditions, hidden by the control actions of the temperature–regulating system.

Shivering as a means of maintaining body temperature within a normal range is often blamed for having a low efficiency, because the movements of the limbs are supposed to disturb the insulating air layer and increase convective heat loss [290]. However, such studies were usually performed in anaesthetized lying animals, whose tremor is rather coarse. In standing goats, the shivering–induced increase in heat loss did not exceed 20% of extra HP [334]. A more valid argument against shivering, and in favour of non–shivering thermogenesis, is, of course, that shivering could interfere with intentional movements and locomotion.

5.6
Non–Shivering Thermogenesis in Brown Adipose Tissue

Many small mammals, with the exception of monotremes and marsupials [112], possess brown adipose tissue (BAT): a specialized organ whose primary product is heat. The BAT fraction of body mass is at most of the order of a few percent. However, its contribution to the HP of a cold exposed animal can exceed 30% and is termed non–shivering thermogenesis (NST) or, to distinguish it from resting HP, facultative NST. In cold–adapted rat, the mass–specific HP of activated BAT was 240 $mW \cdot g^{-1}$ of tissue, while it was 4 $mW \cdot g^{-1}$ in the liver [134]. Thus, BAT behaves like a small but highly effective oven.

Its brown adipocytes are densely innervated by **sympathetic fibres**, whose signals originate in the ventromedial hypothalamus [88,504]. Within minutes upon cold exposure, noradrenaline is released and bound primarily to ß–adrenergic receptors of the plasma membrane. The binding triggers a cascade of intracellular messenger signals which result in two effects. The first is increased hydrolysis of stored lipids, supplying free fatty acids as fuel for the mitochondria. The second is that the rate of oxidation of free fatty acids in mitochondria multiplies [507].

In conventional mitochondria of the body, and in those of non–activated brown adipocytes, the rates of oxidation, and hence HP, are closely linked to the rate of ATP synthesis. The coupling is established by protons, which are extruded across the inner mitochondrial membrane in the process of oxidation and return via the ATP synthase. In **activated brown adipocyte mitochondria**, however, protons can bypass the ATP synthase and enter the mitochondrial matrix via an uncoupling protein (UCP). This has two effects: first, the cycling of protons, which is the rate–limiting factor of oxidation and HP, is greatly accelerated; second, because oxidation is uncoupled from ATP synthesis, the energy released during oxidation is not conserved as ATP but dissipated immediately as heat [372].

UCP exists in two states: no passage for protons in the non–activated condition vs. free passage during NST. The passage is blocked by nucleotides (ATP, ADP and others) binding to UCP. Thus, on activation of NST, an intracellular signal, derived from adrenoreceptors of the plasma membrane, releases nucleotides from UCP and opens the proton passage [237,407]. The role of BAT may extend beyond the production of heat, at least in some small rodents. In cold–adapted rat, activation of NST occurs in parallel with increased production, in BAT, of triiodothyronine (T3) from thyroxine. T3 is locally important for the production of UCP, but is exported also to the rest of the body [458]. Higher systemic levels of T3 are associated with higher HP in tissues other than BAT [221].

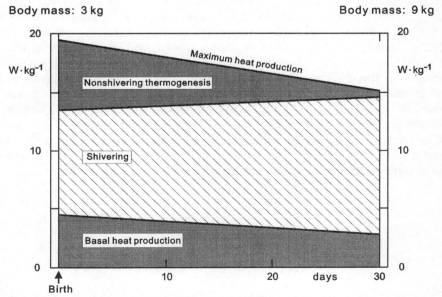

Fig. 5.4. Maximum heat production during severe cold exposure and its fractions (basal heat production, heat production by shivering, and heat production by non–shivering thermogenesis) in lambs from birth to 1 month of age. (After data from [4], with permission)

BAT is present throughout the lifetime of hibernators, to whom it is of vital importance during rewarming, and other small mammalian species. Its growth and involution vary with environmental temperature, photoperiod and the availability and composition of food [219]. In larger species, including humans, however, the presence of BAT is restricted to the perinatal period. With increasing age, the amount of BAT becomes smaller, and NST is gradually replaced by shivering (Fig. 5.4). As a general rule, BAT as a source of heat is negligible in animals of more than 10 kg body mass [192].

BAT is found at various sites within the body, preferentially in or near the thoracic and abdominal cavities. A primary site is the interscapular region. It is richly vascularized so that the heat produced by the tissue is rapidly distributed in the body core. In newborn lambs during severe cold exposure (Fig. 5.4), 20% of cardiac output passed through BAT which amounted to only 1.5% of body mass [3]. Animals possessing BAT also employ shivering during severe cold exposure. Both effector mechanisms are sequentially called upon: NST is activated by milder cold loads, and shivering is added during more severe ones [56].

5.7
Cold–Induced Non–Shivering Thermogenesis in Other Tissues

Recent findings in hibernators suggest that UCP is not unique to brown adipocytes. Homologues occur also in other tissues, potentially contributing to NST [55]. Furthermore, in all animals responding to cold by shivering or NST, the higher demand for oxygen poses an additional work load to respiratory muscles and the heart, which is clearly the source of some facultative non–shivering thermogenesis. Blood vessels of non–contracting muscles respond to noradrenaline by constriction, which cause a detectable increase in HP of the vessels' smooth muscles [94]. Also, the levels of a number of hormones such as epinephrine, glucagon, insulin and others are affected by acute cold exposure, and it is not unlikely that these hormones contribute, to a hitherto unknown degree, to higher HP in several tissues [254]. However, the importance of other tissues and organs for the metabolic response to cold is certainly secondary to that of BAT and skeletal muscle.

6 Physics of Heat Exchange with the Environment

The four modes of heat exchange between an animal and its terrestrial environment are conduction, convection, radiation and evaporation. The rates of heat transfer (**watt**) by all modes are proportional to the area at which the transfer takes place. Thus, as far as the skin is concerned, the rates are expressed per unit surface area ($\mathbf{W \cdot m^{-2}}$). In a first approximation, the rates of heat transfer by conduction, convection and radiation are proportional to the temperature difference between the surface and the environment; it follows that the physical features of these three processes are described per unit temperature difference ($\mathbf{W \cdot m^{-2} \cdot {}^{\circ}C^{-1}}$). In the case of evaporation on the skin, the water vapour pressure difference (**kPa**) between the body surface and the ambient air substitutes for the temperature difference.

6.1 Conduction

Conductive heat transfer is the process by which the energy of random motion of molecules of a solid or fluid at higher temperature is transferred to molecules of a solid or fluid at lower temperature. The specific rate–determining factor is the **thermal conductivity** of the solids or fluids involved. Common experience illustrates the point: one can take, with bare hands, a piece of wood from a hot chamber at 100 °C without harming the fingers. Because the apparent thermal conductivity of blood–perfused skin is higher than of wood, heat is removed from the fingers at a larger rate than supplied from the hotter part of the piece to the points of contact (the qualification "apparent" accounts for blood flow adding an internal convective component to the thermal conductivity of non–perfused skin). As can be seen in Table 6.1, it is not advisable to repeat the experiment with copper.

Table 6.1. Thermal conductivity of some materials. The unit is ($\mathrm{W \cdot m^{-1} \cdot {}^{\circ}C^{-1}}$): Watt transferred over a distance of 1 m if the temperature difference between both ends is 1 °C

• Air	0.025
• Fur	0.038
• Helium	0.150
• Fat, excised	0.200
• Skin, excised	0.400
• Muscle, excised	0.500
• Water	0.600
• Concrete	0.900
• Sandy soil, 30% water content	2
• Copper	380

If conduction is to be an appreciable component of an animal's overall heat exchange, the conductivity of the contact material must be relatively large, its temperature must be clearly different from the surface temperature of the animal, and the area of contact must be a relevant fraction of the animal's surface. This is obviously not the case in standing animals, and also in sitting animals the limbs are usually folded underneath the body so that the thorax and the abdomen are partially kept away from the ground. However, conductive heat loss matters when an animal is lying on wet soil or, possibly in husbandry, on concrete.

Conduction is energy transfer and involves no mass transfer. Thus, in the case of conduction between a solid and a fluid, that is, an animal and its ambient air or water, the definition requires that **the fluid is still**. In bare–skinned animals, this condition is met only by an infinitesimally thin layer next to the skin. Further away from the surface, air or water are moving so that the heat transfer is essentially determined by convection.

6.2
Convection

Convective heat loss involves the mass transfer of an ambient medium relative to the body and is treated in two parts. The first refers to natural convection, which occurs when a warmer body is placed in cooler still air or water. The second is forced convection and takes place when the medium around the body is kept moving by external forces such as in currents of wind or water, or when the body moves itself through a still medium.

6.2.1
Natural Convection

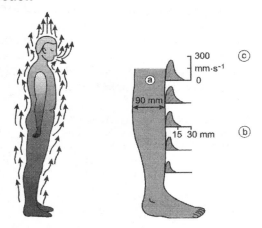

Fig. 6.1. Air flow by natural convection. *Left* An upward–streaming sheath of warm air develops around a standing subject because the air becomes more buoyant by taking up heat from the warmer skin. *Right* Velocity and thickness profiles of the air stream at a leg. Scaling: *a* diameter of the leg; *b* distance from the surface; *c* velocity. (After [90], with permission)

A man is standing in a closed room (Fig. 6.1). Air temperature is 25 °C, and mean skin temperature is 33 °C. The cooler air in contact with the warmer feet takes up heat by conduction, and becomes more buoyant. The same occurs at all other skin sites. The result is an–upward streaming **sheath of warmed air**. This is the mechanism of natural convection, by which heat is carried away from the body. At any skin site, the local heat loss depends on the temperature gradient, and the thickness and velocity of the sheath. However, an overall **natural convective heat loss coefficient** can be estimated for the body as a whole. It describes the heat loss in $W \cdot m^{-2}$ surface area and °C temperature difference between skin and air, and is of the order of 3 $W \cdot m^{-2} \cdot °C^{-1}$ for human beings [90,336]. Thus, in the case of the man with 33 °C skin temperature in 25 °C air temperature, the heat loss by natural convection is 24 $W \cdot m^{-2}$, which is about one half of his resting heat production.

6.2.2
Forced Convection

It is intuitively understood that, for a given temperature difference between body surface and air, the heat loss by convection increases with rising **air velocity**: a fresh breeze can make one feel cold even on a sunny day and at an otherwise comfortable air temperature. It is less evident that, at a given air velocity and temperature difference between surface and air, the **heat loss per unit of surface area** is large in a small animal, and **decreases with increasing body mass** in animals of similar shape (Fig. 6.2). The process of convection is governed by the laws of fluid dynamics which involve several physical properties of the medium, and shape and size of the body. A detailed treatment of this complex subject is beyond the scope of this text and can be found elsewhere [41,138,287,336,349]. However, a general description may help to understand the essence of the problem.

Consider a cylinder, the usual analogy of an animal, exposed to a homogeneous air stream of a given velocity. When hitting the cylinder, the streamlines follow its curvature in almost parallel order. However, the air velocity is proportional to the distance from the cylinder: the closer a streamline is to the surface, the lower is its velocity. Actually, the velocity of an infinitesimally thin **boundary layer** of air next to the surface is zero.

In **laminar flow**, the thickness of the air sheath within which the velocity is affected by the presence of the cylinder, increases with its diameter: the thicker the cylinder, the thicker is the layer of air whose velocity is reduced, and the lower is the rate of heat loss by convection. Conversely, a thin cylinder distorts the air stream less than a thick one, and even at closer distances from the surface, the air velocity is affected only to a minor extent. It is for this reason that at a given air velocity, a small animal (or a smaller part of a larger animal) loses more heat, per units of temperature difference and surface area, than a large one.

The lines in Fig. 6.2 were calculated for different diameters of cylinders with smooth surfaces, representing animals of different size [41]. This is clearly a very substantial simplification: animals have legs and their surfaces are far from being smooth. Still, at least in human subjects reasonable agreement exists between theoretically expected and experimentally determined data [90]. The man with 33 °C skin temperature in 25 °C air temperature lost 24 W·m⁻² by natural convection in still air (Fig. 6.1). If he exposed himself to a gentle breeze of 5 m·s⁻¹, the heat loss by forced convection would amount to nearly 150 W·m⁻², which equals three times his resting heat production and presents a sizeable cold load. On the other hand, the heat loss of a horse racing at 15 m·s⁻¹ in still air is greatly facilitated.

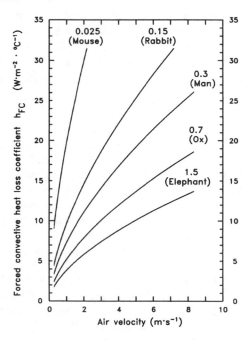

Fig. 6.2. Forced convective heat loss coefficients hFC (W·m⁻² surface area and °C temperature difference between surface and air) as a function of air velocity, calculated for cylinders of different diameter (*m*), representing different species. (After [41], with permission)

6.2.3
The Heat–Transferring Medium

The foregoing sections dealt with convective heat loss in air and, for the sake of simplicity, the physical properties of air and their relevance to heat transfer were not considered. However, humans engage in deep–sea diving, down to several hundred m below sea level and for extended periods of time. Under these high–pressure conditions (30 bar or more), nitrogen as the inert component of the breathing gas is replaced by **helium**. A typical environment in a bell at a depth of 300 m consists of 1.3% oxygen and 98.7% helium at a pressure of 31 bar. In this situation, it is relevant that the convective heat loss coefficient (hFC) depends not only on the velocity of the medium and the shape and size of the body, but also on the physical properties of the ambient medium such as the **thermal conductivity, specific heat, density and viscosity**. The thermal conductivity and specific heat of helium are considerably greater than of air. The consequence is that the h_{FC} in a helox mixture at 31 bar is 15 times greater than in air at sea level [231].

Similar considerations apply to **water**. The thermal conductivity and specific heat of water are nearly 20 and 4 times greater than of air, respectively, and the pattern of flow is different [90]. Thus, the h_{FC} in water is approximately 100–200 times greater than in air [138]. In humans swimming at a speed of 0.5 $m \cdot s^{-1}$, the h_{FC} was 580 $W \cdot m^{-2} \cdot {}^{\circ}C^{-1}$ [360]. No such data are available for other species. However, one conclusion applies to all mammals: except at very high levels of activity, the gradient between skin and water temperatures must be kept small, of the order of a few tenths of a °C, if the heat loss is to be balanced by heat production.

6.3
Radiation

Electromagnetic radiation is emitted and absorbed by all solids, liquids and gases. The energy emitted by a plane surface of a black body or full radiator is proportional to the fourth power of the absolute surface temperature:

$$\text{thermal radiant exitance } [W \cdot m^{-2}] \ = \ \sigma \cdot T^4 \qquad (6.1)$$

where the Stefan–Boltzmann constant σ is 5.67×10^{-8} $[W \cdot m^{-2} \cdot K^{-4}]$. Thus the radiant exitance of the sun (T = 6000 K) is approximately 74 000 $kW \cdot m^{-2}$. However, owing to the distance, not more than 1360 $W \cdot m^{-2}$ arrive at the outer atmosphere of the earth (solar constant). The same calculation for a body surface of 37 °C (T = 310 K) yields a thermal radiant exitance of 525 $W \cdot m^{-2}$. Also the spectral distribution of thermal radiation is determined by the absolute temperature of the surface (Wien's law). The wavelength (lambda) of maximum energy is:

$$\text{lambda}_{max} [\mu m] \ = \ 2897/T \qquad (6.2)$$

Lambda$_{max}$ is 9.3 μm for the radiation emitted by a 37 °C surface, and 0.48 μm at the surface of the sun. However, the spectral distribution of solar radiation arriving at the surface of the earth deviates from the extraterrestrial solar spectrum. Owing to complex and variable effects of scattering and absorption by atmospheric gases, it is skewed towards the long end [349]. With regard to living matter, the wavebands listed in Table 6.2 are of particular importance.

Table 6.2. Biologically relevant wavebands of electromagnetic radiation

•	Ultraviolet	0.25 - 0.40 μm
•	Visible	0.40 - 0.75 μm
•	Short–wave infrared	1 - 3 μm
•	Long–wave infrared	3 - 100 μm

6.3.1
Long–Wave Infrared

The temperatures prevailing at the earth's surface confine the radiation emitted by natural objects to 3–100 μm, and within this range any natural object behaves nearly like a full radiator, which is defined as a perfect absorber of all incoming radiation and an equally perfect emitter. Thus the **net long–wave radiant heat exchange** between two objects of different temperatures (T_1, T_2) can be computed by a modification of Eq. 6.1:

$$\text{radiant flux } [\text{W·m}^{-2}] \; = \; \sigma \cdot (T_1^4 - T_2^4) \tag{6.3}$$

The fourth powers in Eq. 6.3 make calculations somewhat tedious. However, in many natural situations the temperature differences are small, and the relationship can be linearized so that the net radiant flux is calculated from the first power temperature difference between two surfaces [503]. In a typical indoor situation (wall temperature 20 °C, animal surface temperature 30 °C), the net flux, or heat loss by radiation, is 60 W·m^{-2}, that is, the **radiative heat loss coefficient** (h_R) is of the order of 6 W·m^{-2}·°C^{-1} [41].

6.3.2
Solar Radiation

Not all of solar radiation arriving at the outer atmosphere of the earth reaches the ground. However, the rest is of sufficient magnitude to have a considerable impact on animals. Figure 6.3 shows, for a location near the equator and 1500 m above sea level, the radiant heat load on the upward– and downward–facing surfaces of a flat horizontal plate [126]. On the way through the atmosphere, part of the solar radiation is scattered by gases, back to space or down to the ground. The second fraction plus the direct solar radiation arriving at the surface are combined in the term **direct and diffuse short–wave solar radiation** (S+s). Both hit the top surface of the plate, while its bottom surface is in the shade. The resulting heat load per unit of total surface area (top and bottom) reaches 480 W·m^{-2} at noon. Its energy is contained in the waveband from ultraviolet to short–wave infrared which is subject to reflection. Thus, part of the direct and diffuse radiation, which misses the top surface, is reradiated from the ground to the downward–facing surface of the plate. Per unit of total surface area, the maximum load attributable to **reflected short–wave** radiation (Sr+sr) amounts to 150 W·m^{-2}.

That part of the direct and diffuse radiation which arrives at the ground and is not reflected heats its surface and gives rise to **long–wave infrared** radiation from the **ground** (R_g). Some of it reaches the bottom surface of the plate, and the maximum load attributable to this component is 300 W·m^{-2}. While passing through the atmosphere, about 10% of the energy of the solar beam is absorbed by gases and water vapour. This fraction heats the atmosphere, which then sends part of it, as **long–wave infrared** radiation from the **atmosphere** (R_a), to the earth. The load arising from this component is about 170 W·m^{-2} at noon.

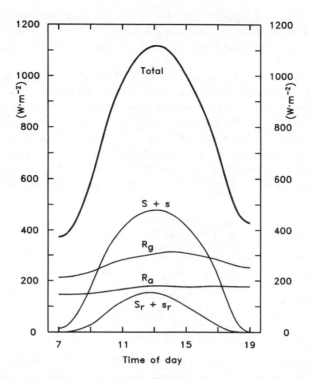

Fig. 6.3. Radiant heat load on the upward and downward facing surfaces of a flat horizontal plate as a function of time of day. The components are (1) direct and diffuse short–wave ($S+s$), (2) reflected short–wave ($Sr+sr$), (3) long–wave from the ground (Rg), (4) long–wave from the atmosphere (Ra). $S+s$ and Ra hit the upward–facing surface of the plate, while $Sr+sr$ and Rg hit the downward–facing surface. All values are given in W·m^{-2} of total surface area (*top* + *bottom* of the plate). (Redrawn and modified from [126])

All components add up to the total radiant heat load of the plate, which reaches a maximum of nearly 1100 W·m^{-2} at noon. This figure is put in the right perspective by the notion that it exceeds the resting heat production of a 100 kg–antelope (80 W·m^{-2}), standing next to the plate, by a factor of 14. For the antelope, however, 1100 W·m^{-2} is the potential radiant heat load, and must not be confused with the heat load of the inner body. One key to survival of large animals in regions of intense solar radiation is insulation by fur, which is dealt with in Chapter 7.

In contrast to long–wave infrared, short–wave radiation is partially reflected. The size of the fraction depends on the colour of the surface. Table 6.3 lists the reflectance from some materials [41]. Thus, in the case of the antelope being a hartebeest, 40% of the incoming short–wave radiation at noon = 250 W·m^{-2} is reflected. Table 6.3 shows that the reflectance of black skin is about half as large as of white skin. However, the disadvantage of the higher heat absorption of black skin is possibly of less significance than the better protection, which is provided by the larger melanin content of the stratum corneum of the skin: melanin absorbs the ultraviolet waveband and protects the deeper layers of the skin against its cancerogenic effects.

The data of Fig. 6.3 do not take into account that the position of an animal relative to the sun, its **solar radiation profile**, changes with time of day even if the animal does not move. The effect is most pronounced in standing human subjects (vertical cylinders): the fraction of the total body surface intercepting the direct solar beam is near 25% at a solar altitude of 10°, but only 5% with the sun directly overhead [509]. The influence of solar altitude is smaller in four–legged animals (horizontal cylinders), while the effect of changing the position, from side to head presentation, at low solar altitudes is larger than in standing human subjects [89].

Table 6.3. Reflectance of solar radiation from some materials. Reflectance is the ratio of the radiant flux reflected by a surface or medium to the incident flux. (Based on data from [41], with permission)

• Short grass	0.24
• Dry sand	0.40
• Black human skin	0.18
• White human skin	0.35
• Hartebeest coat	0.40

The considerations above dealt with the situation of animals living in regions of intense solar radiation. However, even in temperate climates and under overcast skies, the radiation balance of an animal in an outdoor environment is usually positive during the day, because its long–wave radiant loss is smaller than the sum of the various gains. It is only during the night that the balance becomes negative, when the skies are clear and the temperature is low [41].

6.4
Evaporation

Cooling by evaporation results from the fact that the conversion of water from liquid to vapour is an endothermic process. The amount of heat involved is termed the latent heat of vaporization and is, depending on temperature, approximately $2.4 \text{ kJ} \cdot \text{g}^{-1}$ of water. The biological significance of the figure becomes apparent by relating it to the basal heat production of an adult human subject ($40 \text{ W} \cdot \text{m}^{-2}$):

$$1 \text{ g H}_2\text{O} \cdot \text{min}^{-1} \approx 2.4 \text{ kJ} \cdot \text{min}^{-1} = 40 \text{ J} \cdot \text{s}^{-1} = 40 \text{ W} \qquad (6.4)$$

Thus, the evaporation of $2 \text{ g} \cdot \text{min}^{-1}$ is sufficient to dissipate the heat generated by the basal metabolic rate of a human subject with a body surface area of 2 m^2.

6.4.1
Sweating

The maximum sweat rate of humans is of the order of 10 to 15 $g \cdot min^{-1} \cdot m^{-2}$, and horses do even better [226], making evident that sweat secretion is a powerful mechanism of temperature regulation. However, the sweat must evaporate, and to do so, the **water vapour pressure** (e, [kPa]) on the skin must be greater than that of the surrounding air. An important term in this context is the maximum vapour pressure, which can be generated by water. It is termed saturation vapour pressure and increases exponentially with temperature [138]. Air having its maximum water vapour pressure at a given temperature has also a relative humidity of 100%. **Relative humidity** is the ratio of a given water vapour pressure to the vapour pressure of saturated air at the same temperature, expressed as a percentage.

The 100% line in Fig. 6.4 presents the saturation vapour pressure of air in the range 20–40 °C. As mentioned before, the **rate of evaporation is proportional to the difference between the vapour pressures of the air and on the skin**. The maximum water vapour pressure on a fully wetted skin is the saturation vapour pressure at skin temperature. Therefore, if evaporation is to be sustained against fully saturated air, the skin temperature must be higher than the air temperature. Figure 6.4 illustrates that heat dissipation (and life) in saturated air is difficult when the air temperature approaches 37 °C. In contrast, air of low relative humidity permits sufficient rates of evaporative heat loss even at very high temperature.

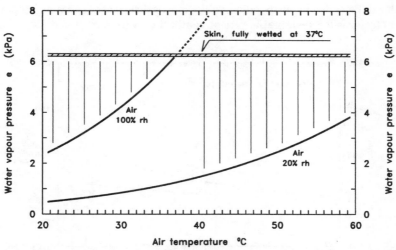

Fig. 6.4. Water vapour pressure (*e*) of air at different relative humidity (*rh*) plotted vs. temperature. Heat loss by evaporation requires that the water vapour pressure of the surrounding air is lower than on the skin, where it is 6.3 kPa at maximum sweat secretion and 37 °C. The heat loss is zero, if the air is fully saturated at this temperature (rh 100%), while drier air (rh 20%) permits substantial heat dissipation even at a temperature of 60 °C. The *vertical lines* show the differences between e of air and maximum e on skin.

In order to convert the differences between the vapour pressures of skin and air (e, kPa) into rates of evaporative heat loss (W·m^{-2}), a proportionality factor is required. This is the **evaporative heat transfer coefficient** (h$_E$), which is of the order of 40 W·m^{-2}·kPa^{-1} in still air [287]. However, the rate of evaporation is not only determined by the difference between the vapour pressures. It is intuitively comprehensible that wind increases evaporation. The rate of evaporative heat loss per unit of vapour difference (h$_E$) is influenced by air velocity and shape and size of the body in the same way as the rate of forced convective heat loss per unit of temperature difference (h$_{FC}$) [326]. Both coefficients bear a fixed relation [287]:

$$h_E \ [W \cdot m^{-2} \cdot kPa^{-1}] / h_{FC} \ [W \cdot m^{-2} \cdot {}^{\circ}C^{-1}] \ \approx \ 15 \qquad (6.5)$$

6.4.2
Panting

Many species have no functional sweat glands and rely on panting to increase evaporative heat loss. On inhalation, air flows over the wet surfaces of the respiratory tract, and the exchange of heat and water can be so efficient that the exhaled air is saturated at internal body temperature. The rate of respiratory evaporative heat loss depends mainly on the **water vapour pressure of the inhaled air** and the **rate of ventilation** (Fig. 6.5). Thus, the maximum heat loss by panting depends on two factors. The physical limit is set by the environment: as for sweating, increasing heat and humidity of the inhaled air curtail the amount of water which can be added by evaporation on the mucosal surfaces. The limiting physiological factor is the **sustainable rate of ventilation** (Chapters 11 and 15).

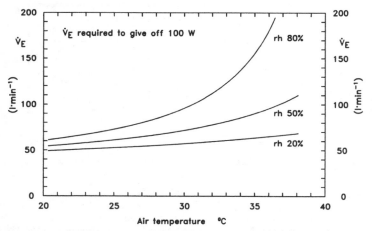

Fig. 6.5. Respiratory minute volume (*VdotE*) required to dissipate 100 W at different relative humidity (*rh*) plotted vs. air temperature. It is assumed that the exhaled air is fully saturated at 38 °C temperature. A fixed rate of respiratory heat loss requires the respiratory minute volume to increase, if temperature and relative humidity of the inhaled air increase

7 External and Internal Insulation

Reindeer and other larger arctic mammals maintain the temperature of the body core (T_{core}) at about 38 °C even if air temperature (T_{air}) is as low as −30 °C [273,448]. It is clear from the laws governing the radiant and convective heat transfer that a temperature gradient of nearly 70 °C implies a huge potential heat loss; and yet, these animals do not shiver and are still in their thermoneutral zone (Fig. 17.5), the range of ambient temperatures in which regulation is achieved entirely by the control of dry heat loss. The key to such perfect adaptation is the insulation of the body core. In analogy to Ohm's law, insulation can be considered a resistance to the ease with which heat leaves or enters the core:

$$\text{current} = \frac{\text{voltage}}{\text{resistance}} \qquad\qquad \text{heat flow} = \frac{\text{temperature gradient}}{\text{insulation}}$$

Rearranging the equation yields insulation:

$$\text{insulation} = \frac{\text{temperature gradient}}{\text{heat flow}} \qquad \left[\frac{°C}{W \cdot m^{-2}} \right] \qquad (7.1)$$

The temperature gradient is the difference between the inner and outer surfaces of the insulating layer. Thermal insulation is expressed per unit area ($°C \cdot m^2 \cdot W^{-1}$). A more comprehensible term is the **thermal conductance (C)** ($W \cdot m^{-2} \cdot °C^{-1}$), the inverse of insulation: watt per unit area, passing a thermal resistance across which a temperature gradient of 1 °C is maintained:

$$\text{conductance} = \frac{\text{heat flow}}{\text{temperature gradient}} \qquad \left[\frac{W \cdot m^{-2}}{°C} \right] \qquad (7.2)$$

The insulation between the body core and its environment consists of two components which, as resistances, are in series and additive. One is subcutaneous fat and shell tissue, and the other is fur. The extent to which both components are utilized varies between orders and species. Most terrestrial mammals living in cold climates rely primarily on fur so that skin temperature (T_{skin}) at least at the trunk is relatively independent of T_{air} (Fig. 17.5). Arctic swine with bare skin, however, use internal insulation by subcutaneous fat: at −10 °C T_{air}, T_{skin} at the trunk fell to

+9 °C and was tolerated without signs of thermal discomfort [248]. Both modes of insulation, external and internal, have their pros and cons, and are, together with appropriate behaviour, powerful means of coping with extreme environments.

7.1
External Insulation: Fur

The insulation provided by fur (or clothing) is of the same order of magnitude as that provided by still air. In fact, fur does not insulate by some peculiar properties of hairs but by maintaining a **shell of trapped air** around the body through which heat is transferred primarily by conduction [176]. The insulating value of the shell depends in the first place on its thickness. Figure 7.1 shows C across layers of different thickness. The line for fur is based on measurements by Scholander et al. in arctic terrestrial mammals between 0.03 and 500 kg [249,450].

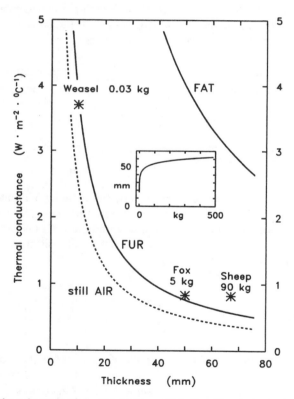

Fig. 7.1. Thermal conductance of air, fur and fat vs. thickness of insulating layer. The conductance of fur (*lower solid line*) is of similar magnitude as that of still air (*interrupted line*), and much lower than that of fat. *Inset* Fur thickness plotted vs. body mass in arctic terrestrial mammals, increasing between 0.03 and 5 kg, and levelling off at larger body mass. (After [450], with permission)

Apparently, a 30–g weasel is not helped very much by its 10 mm of fur, but it cannot wear more without being hindered when moving, and is confined to burrows for most of the winter time. The **thickness of fur** increases steeply with body mass, and reaches 50 mm in white polar fox, a species of 5 kg mass. At this thickness, a sufficient reduction in heat flow appears to be reached, and the additional 15 mm of Dall sheep do not further improve insulation, at least in still air.

The difference between C of still air and fur points to the fact that the air trapped between hairs is never perfectly still. Any temperature gradient across a layer of air or fur induces natural convection, and the line for the heat transfer across still air, which was calculated from its thermal conductivity (Table 6.1), is purely theoretical. Another aspect is that an animal can increase the fur thickness by **piloerection**. Actually, a doubling of the thickness was found in calves acutely exposed to cold stress. The reduction in C, however, was smaller than expected, presumably because of larger convection through the less dense fur [41].

A major problem to larger animals in a cold environment is exposure to **wind**. The moving air penetrates the fur so that the shell of still air becomes thinner. The result is forced convection in the outer layers of the fur and increasing C. The magnitude of the effect depends on air velocity and the angle at which the wind hits the surface. The worst case is a wind blowing perpendicular to the skin plane, because the fur is simultaneously compressed and penetrated.

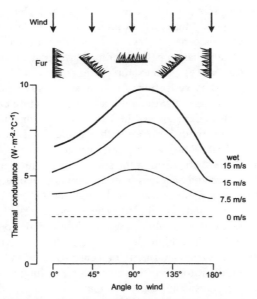

Fig. 7.2. Thermal conductance of fur at various air velocities plotted vs. angle of skin surface to wind direction. At all angles, the conductance increases with air velocity. At a given velocity, the conductance depends on the angle at which the wind hits the surface. If the fur is slightly wet (*upper heavy curve*), the conductance increases further. Data from newborn caribou with average fur thickness of 11 mm. (After [304], with permission)

In the experiments shown in Fig. 7.2, an air velocity of 15 m·s⁻¹ tripled the heat flow relative to that in still air. The situation deteriorates further when wind is combined with **rain**. The upper curve in Fig. 7.2 shows the increase in C caused by water droplets in the fur (as little as 1.7% by volume). Apparently, newborn caribou are not well equipped to endure cold rain. However, fur can retain its insulating properties even in water. A prime example is semiaquatic fur seal whose external insulation includes 20 mm of dense water–repellent underfur, under which 35 °C T_{skin} was maintained in water at 10 °C [252].

7.1.1
Fur and Solar Radiation

The protective value of fur against cold is commonplace even to inhabitants of temperate climates. Less obvious is that fur is also a key to survival of medium–sized animals in regions of intense solar radiation. On sunny days near the equator, the **potential radiant heat load**, that is the sum of all direct and indirect, short– and long–wave radiation incident on the body surface of an animal, can be as great as 1100 W·m⁻² (Fig. 6.3). This exceeds the rates of resting heat production in all animals by several times. Obviously, a radiant flux of such magnitude must be prevented from entering the body, because returning it to the environment by evaporation of water would possibly exceed the capacity of the sweating or panting mechanisms, and certainly put a severe strain on the usually tight water budget in arid regions. Mice retreat to cool burrows, and elephants have a very favourable ratio of surface area to body mass which limits the rate of heat inflow per unit of body mass. For intermediate species, external insulation by fur is part of the solution to the problem. The radiation balance of the hartebeest, an African antelope of 100–150 kg body mass, serves as an example (Fig. 7.3).

A minor role is played by **reflection**. At noon, short–wave radiation amounts to more than one half of the potential radiant load. The reflection coefficient (ratio of reflected to incident flux) depends on the colour of the fur and is around 0.4 in hartebeest, which is light reddish brown with a white tip at the end of each hair. Thus, approximately 250 W·m⁻² were simply reflected to the environment. The remaining 850 W·m⁻² were absorbed at the surface of the fur. However, its low C, being so effective in reducing heat loss in arctic animals, is of equal importance to desert animals: it reduces the heat transfer through the fur so that, to state it simply, absorbed heat accumulates at the surface. In hartebeest, the fur surface assumed a temperature of 46 °C, while T_{skin} underneath the dense fur of 8 mm thickness was just 40 °C. The high surface temperature generated a **long–wave radiant flux from fur to environment** of nearly 600 W·m⁻² (and in addition, would increase the gradient for convective heat loss, as long as fur temperature is higher than air temperature). The result was that only a minor fraction of the potential radiant heat load actually flowed into the body [126].

A domestic species performs even better. Merino sheep in arid regions of Australia, South Africa and South America grow fleece of up to 80 mm thickness. At intense solar radiation, the outer fleece temperature reached 85 °C while T_{skin}

remained at 42 °C [315]. In summary, reflection at the surface and resistance to heat flow from the surface to the skin are the decisive features of fur which permit medium–sized animals to thrive in regions of intense solar radiation.

If that is so, one may wonder why desert animals are not white, which would give them more benefits from reflection. However, the **colour of the fur** was certainly also selected by non–thermal factors such as its ability to hide an animal. This does not explain why the black Bedouin goat of the Negev desert thrives in open defiance of the advantage of light colours. The climate of the Negev is characterized by high solar radiation in summer and very cold days in winter. One hypothesis is that the benefit of a larger solar heat gain on winter days (when food supply is scarce) may outweigh the disadvantage in summer [117]. Another possibility is that the smaller reflection is more than compensated by a relatively less deep **penetration** of non–reflected radiation in black fur [141].

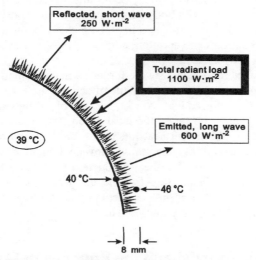

Fig. 7.3. Radiation balance of fur during intense solar radiation. The total radiant load is the sum of all direct and indirect, short– and long–wave radiation incident on the body surface. A minor part was reflected by the light fur. The major part of the load was absorbed and heated the surface to 46 °C because the low conductance of fur impeded the heat transfer to the skin. The high surface temperature was the source of long–wave infrared radiation by which more than one half of the total load was returned to the environment. *Dots* Local temperatures at the surfaces of fur and skin; *circle* Internal body temperature. (Based on data from hartebeest [126])

7.1.2
Fur and Exercise

A general problem of fur as an insulating barrier is that it represents nearly a **fixed resistance**. Animals may adapt to the annual cycle of climates by shedding their fur and replacing it by another more suitable one for the forthcoming season, and most species do: the process is controlled by photoperiod and environmental temperature [524]. However, this is, of course, a long–term adjustment. As mentioned earlier, quick variations of the resistance to heat transfer across the fur can be achieved by altering its thickness (hair erect or flat), but the mechanism is of very limited capacity and certainly not potent enough to deal with the problem of heat loss during exercise. All African ungulates are prey animals, and the heat production during a chase can be expected to exceed ten times the resting level. In such situations, thick fur, while providing excellent insulation against solar radiation, is a nuisance because it impedes the heat loss by convection and evaporation from the body surface.

The problem increases with body size: the larger the animal, the smaller is the surface area per unit of heat–generating body mass which is available to dissipate the waste heat of exercise. Thus, to make the most efficient use of the relatively smaller surface area, the fur should be as thin as can be tolerated for the sake of protection against solar radiation. It may be for this reason that the fur thickness in African ungulates changes with body mass in a direction opposite to that in arctic terrestrial mammals. The fur thickness in 500–kg eland is just 1 mm. A swift antelope like springbok, prone to sprinting at great speeds, has even thinner fur than predicted from its body mass [229].

The problem of exercise is less severe for terrestrial mammals living in cold regions. Species like reindeer, in which bouts of activity are a frequently occurring part of life, have thinly furred legs. At rest, leg skin blood flow and local heat loss are low. During running, however, the local blood flow multiplies and results in enhanced convective and radiant heat loss. Thus, the legs and other thinly furred regions such as the ears are like thermal windows, which give flexibility to adjust the heat loss to different levels of heat production [131,351]. This topic is dealt with in more detail in Chap. 10. Also, the efficiency of respiratory heat loss mechanisms is greatly enhanced in a cold environment, because the water vapour pressure of the inhaled air is very low (Fig. 6.5).

7.2
Internal Insulation: Fat and the Principle of Core and Shell

The answer of **bare–skinned mammals**, terrestrial or aquatic, to seasonal or permanent cold exposure is a subcutaneous layer of fat or blubber. The insulating value of fat is clearly inferior to that of fur (Fig. 7.1), which explains the sometimes enormous thickness of blubber. In harp seal, 60 mm were necessary to insulate the 37 °C body core from ice water; then T_{skin} was nearly identical to water temperature and the heat loss was minimized [250]. However, insulation by blubber offers two advantages over fur. The first is important to aquatic mammals: blubber is relatively incompressible, and hence far superior to fur in mammals which must dive to depth [314]. The second applies also to terrestrial species: the resistance to heat flow is variable, because the layer of insulating blubber can be bypassed by high blood flow to the skin, permitting a large degree of flexibility in matching heat loss with varying levels of heat production.

7.2.1
Core Temperature vs. Mean Body Temperature

Effective internal insulation does not necessarily require thick layers of subcutaneous fat: the thermal conductivity of any non–perfused tissue is of the same order of magnitude as that of fat (Table 6.1). Thus **all peripheral tissues**, in particular skeletal muscle and skin, can **contribute** substantially **to the insulation** of the body core, provided peripheral blood flow is low. However, the rates of blood flow through muscle and skin are variable, and so is the peripheral insulation. The variation of internal insulation as a response to the thermal state of the body is a unique feature of temperature regulation in animals, and the underlying mechanism is **adjustment of regional blood flow** in all peripheral tissues (Chap. 10).

The result is that the temperature field of the body is actively altered in different environments [24]. The principle is simple and shown in Fig 7.4. In a cool ambience, the warm **core** comprises the cavities of the skull and the trunk, and the trunk core is insulated by a thick **shell**. Low rates of skin and muscle blood flow envelop the body core in layers of low thermal conductivity, whose temperatures approach the temperature of the environment, decreasing gradually from the inner to the outer. The thicker the shell, the closer comes T_{skin} to environmental temperature, and the smaller is the dry heat loss. Upon transition from a cool to a warm environment, however, peripheral blood flow increases, and the core expands from its obligatory component into more superficial layers, reducing the axial and radial temperature gradients also in the limbs. Finally, a large core is surrounded by a thin shell, and heat loss increases [520].

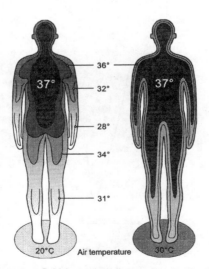

Fig. 7.4. The internal temperature field in cool (*left*) or warm (*right*) environments. The *darkest area* shows the body core, whose temperature is 37 °C in both conditions, and the *gradations of grey* indicate lower tissue temperatures of the shell. The *lines* are isotherms of internal temperature. The core is small during cold exposure, and large in a warm environment. (After [11])

In the simple version of Fig. 7.4, the variable thickness of the **shell serves as a buffer**, when the body is exposed to mild and short–lasting thermal demands, and T_{core} is thought to remain constant. During intense cold exposure, some decrease in T_{core} cannot be prevented, but it is reduced because the shell delivers the major fraction of heat loss to the environment. The importance of this mechanism as a means of dampening changes in T_{core} depends on the capacity of the shell to release or store heat. It can be deduced from contrasting the major change in **mean body temperature** with the minor change in **core temperature** during cooling and rewarming. In Fig. 7.5, mean body temperature, that is, the average temperature of all body tissues, changed by more than 5 °C, while the change in T_{core} was limited to 1.5 °C. Thus, the temperature of the shell must have changed considerably more than mean body temperature.

It is not unlikely that the ability of different species to expand or contract the thermal core is related to differences in body shape. It may be particularly well developed in humans and other long–legged species, in which the limbs present a considerable fraction of body mass and, as a whole, can become part of the shell. However, also in steers the variations of mean body temperature were found to be twice as large as those of T_{core}, when the animals were exposed to a fluctuating thermal environment [328].

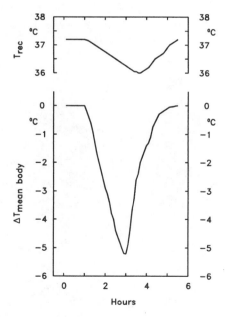

Fig. 7.5. Core temperature (T_{rec}) and change in mean body temperature ($\Delta T_{mean\ body}$) during cooling and warming of a human subject. For calculation of $\Delta T_{mean\ body}$, changes in body heat content were determined from the differences between heat production and heat loss, and converted by assuming a body mass of 75 kg and a specific heat of 3.47 $kJ \cdot kg^{-1} \cdot °C^{-1}$. (After [521])

7.2.2
Internal Conductance

Another way of quantifying the insulating properties of the shell is to determine the internal conductance (C_{int}): the rate of heat flow across the **gradient between core and skin**, per unit of temperature difference. In a resting subject at constant body temperature, C_{int} equals heat production (HP) minus respiratory heat loss (REHL), divided by the difference between T_{core} and mean T_{skin}:

$$C_{int} \left[W \cdot m^{-2} \cdot {}^{\circ}C^{-1} \right] = \frac{HP - REHL}{T_{core} - T_{skin}} \qquad (7.3)$$

Despite some methodological objections concerning the determinations of T_{core} and T_{skin}, C_{int} is a useful measure of the extent to which the heat transfer from the core to the body surface can be modulated. The individual minimum depends of course on the thickness of subcutaneous fat, putting an additional layer of low thermal conductivity between muscle and skin [535].

Interestingly, minimum C_{int} at standardized anthropometric measures is subject to some ethnic variation. Native inhabitants of central Australia used to sleep, without protective clothing and shields against the cold night skies, and C_{int} was nearly 25% lower than in Caucasians exposed to the same conditions [177].

During heat exposure, the increase in peripheral blood flow bypasses convectively the thermal resistances of muscle and subcutaneous fat, and maximum C_{int} in rest was of the order of 25 to 30 $W \cdot m^{-2} \cdot {}^{\circ}C^{-1}$ [342]. Human subjects exercising in a hot environment showed even values of up to 70 $W \cdot m^{-2} \cdot {}^{\circ}C^{-1}$ [42,534]. In this situation, the shell is extremely thin, and the upper limit of C_{int} is essentially defined by the maximum of skin blood flow. Thus, the variability of C_{int}, from the minimum in the cold to the maximum in the heat, is at least six to one.

7.3
Total Conductance

In studies on animals, the minimum conductance is often used as a measure of a species' adaptation to a cold environment. In these cases, it is reasonable to determine the total instead of the internal conductance. The difference is that the gradient between T_{core} and T_{skin} is replaced by the **gradient between T_{core} and T_{air}**. Thus the insulating effect of the fur, in addition to that of the shell tissues, is included. Further usual simplifications are that the subtraction of REHL is omitted and heat flow is given per unit of body mass [total wet conductance, Eq. (7.4)].

$$C_{wet\ total} \left[W \cdot kg^{-1} \cdot {}^{\circ}C^{-1} \right] = \frac{HP}{T_{core} - T_{air}} \qquad (7.4)$$

The minimum of C_{wet} total in different species is inversely related to body mass
[57]. The effect of mass is pronounced in the low range: mass–specific minimum
C_{wet} total of a 50–g animal is nearly three times larger than that of a 500–g animal (Fig. 7.6). Major reasons are the increasing ratio of surface area to body mass,
which implies a smaller scope for establishing an internal body shell (Chap. 10),
and the decrease in fur thickness with decreasing body mass.

The determination in a laboratory is simple: provided a constant T_{core}, C_{wet}
total equals the increase in heat production per °C fall in environmental temperature below the thermoneutral zone [41]. However, the simplicity of determination
corresponds with the limited predictive value for real–life situations. The "true"
conductance of a free–ranging animal depends on posture (effective vs. real surface area), air velocity and radiation exchange, and may be considerably larger or
smaller than that determined in a metabolic cage in a cold room [330].

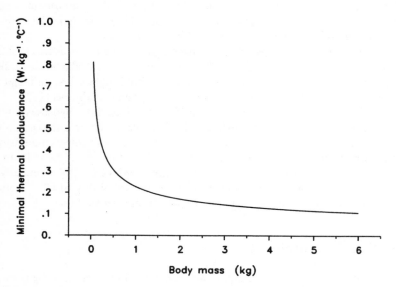

Fig. 7.6. Minimum thermal conductance (C_{wet} total) plotted vs. body mass from 0.05 to 6 kg.
Based on 192 mammalian species. (After [10], with permission)

8 The Temperature Field of the Body Core

8.1
Passive Temperature Deviations

For many aspects of temperature regulation, it is sufficient to treat the temperature of the body core as a uniform entity. However, it is a simplification. The temperature of an organ, and, in fact, of any site within it, depends on the temperature of the arterial blood flowing into the organ, **local heat production** and **local rate of blood flow**. The organs of the body differ in the two last factors, and so do local temperatures, leading to passive thermal gradients within the body. In humans resting in a thermoneutral environment, more than 70% of the heat is produced in the brain, heart, liver and kidneys, whose combined fraction of total body mass is just 8% [11]. These organs are warmer than the remainder of the inner body. Apart from a single modelling study in humans [531], no systematic data are available concerning the temperature distribution in animals. However, it is presumably safe to assume that under normal conditions, that is, rest in thermoneutrality, the mean temperature of any organ in the inner body does not deviate by more than 0.4 °C from arterial blood temperature as the reference. Intraorgan differences can be larger [188], and the temperature of the fetal lamb was reported to exceed, in late pregnancy, that of the ewe by 0.7 °C [299].

The metabolically most important organ outside the cavities of the body is musculature. Its heat production during rest is much lower than that of the inner organs, and this is one reason why the **temperatures in the limbs** are also lower in cold–exposed resting animals. From rest to exercise, however, the metabolic rate of skeletal muscles can increase by a factor of 100 [7]. Consequently, muscle temperature during exercise exceeds blood temperature: in horses running at 90% of their maximum oxygen uptake, muscle temperature was 43.9 °C at a blood temperature of 42.2 °C [226]. High work rates at high body temperature have secondary effects also on the inner temperature field. In a study on humans, blood temperature in a hepatic vein exceeded rectal temperature by 1.5 °C, because liver blood flow was reduced [424].

8.1.1
Where to Measure a "Representative" Body Core Temperature

It is a regular necessity in animal research and practical medicine, to present the complex temperature field of the body core by a single value taken at one site. Ideally, the temperature at this site should be close to the average and follow its changes as fast as possible. For theoretical reasons, the temperature of the **arterial blood as it leaves the heart** is to be preferred: it is the nearest that can be got to a single core temperature which is of definable and general thermoregulatory significance [47]. The temperature of the **mixed venous blood** in the right ventricle or pulmonary artery is an acceptable substitute, except in exercising horses and other high–performance animals, in which heat loss from the lungs produces significant temperature differences between pulmonal and carotid arteries [60,226].

If invasiveness is a matter of concern, **oesophageal temperature** is the best choice in humans and animals [60,246]: the tip of the probe should be positioned at a depth where the left atrium of the heart and the oesophagus are in contact. For unknown reasons, steady–state **rectal temperature** is higher than arterial blood temperature; a further problem is its delayed response in dynamic conditions [44,96]. A still debatable question is whether **tympanic temperature** is a representative core temperature in humans [30,64]. The opponents hold that it is liable to contamination by cool venous blood returning from the scalp and face [61,266]; this problem could be solved by servocontrolled heating of the outer ear [281].

Since the development of radiotelemetry, **abdominal temperature** probes are frequently employed in free–ranging animals, chronobiology and fever research [429,436,515]. Because the position of the probe in the peritoneal cavity is usually unspecified and not necessarily constant over extended periods of time, it is most useful in studies in which thermal gradients within the body core can be neglected.

8.2
Regulated Temperature Deviations

In some animal orders, the temperature of single organs of the body core is temporarily or permanently lower than that of the arterial blood leaving the heart, although neither the specific heat production nor the rate of blood flow differs significantly from those of other organs. The only way by which this can be accomplished is **selective precooling of the arterial blood** before it enters the organ. It results from special vascular arrangements: the arteries supplying blood to the organ are in close contact with veins carrying cool blood from the body surface, and heat is transferred from the arterial to the venous side.

8.2.1
Selective Brain Cooling

When artiodactyls – an order comprising pigs, camels, deer, sheep, cattle, antelopes and others – develop hyperthermia, the temperature of the brain rises less than the temperature of the rest of the body core [13,338]. The phenomenon is termed selective brain cooling (SBC), and its morphological basis is an efficient **countercurrent heat exchanger** situated below the rostral brain stem (Fig. 8.2). The arterial blood destined for the brain passes through the carotid rete, a network of hundreds of small arteries, arising from branches of the carotids and after 10–15 mm joining again to enter the circle of Willis. The rete is fully embedded in the cavernous sinus, which receives cool venous blood returning from the mucosal surfaces of the nose. The system is optimized by the large total surface area of the rete arteries, and can lower the temperature of the arterial blood, and hence brain temperature (Tbrain), by 2 °C or more (Fig. 8.1).

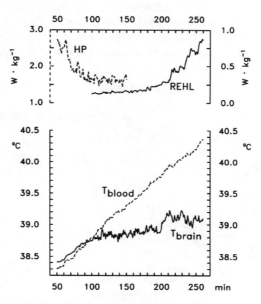

Fig 8.1. Selective brain cooling in a resting goat. The temperatures of the arterial blood (*Tblood*) and the trunk core were raised from 38.4 to 40.4 °C by heat exchangers in the blood stream. In the first phase, body temperature was below normal: heat production (*HP, interrupted line, left ordinate*) was elevated by shivering, and brain temperature (*Tbrain*) exceeded Tblood. At min 100, Tbrain passed a threshold at 38.8 °C and uncoupled from Tblood. With progressive hyperthermia, respiratory evaporative heat loss (*REHL, solid line, right ordinate*) increased, and Tbrain was finally 1.3 °C lower than Tblood. (After [297])

An important feature of the system is that the carotid rete heat exchange, and hence SBC, is not mandatory. There are **two routes** available to the cool venous blood returning from the nasal mucosa (Fig. 8.2). One is via the angularis oculi vein; if this route is taken, the arterial blood flowing to the brain is cooled and SBC ensues. The other route goes via the facial vein straight back to the jugular vein, in which case the cavernous sinus is bypassed and the brain is not cooled.

Segments of the vessels downstream of the division of the dorsal nasal vein contain muscular sphincters, which are richly innervated by sympathetic fibres. The adrenoceptors of the angularis oculi sphincter are of the α–type and the sphincter normally is relaxed, whereas the facial vein has ß–receptors and is normally contracted [272]. Thus, low **sympathetic activity** in the fibres to the veins directs cool venous blood to the cavernous sinus, and the degree of SBC is large.

Conversely, higher sympathetic activity constricts the angularis oculi sphincter, dilates the facial sphincter, and directs the cool venous blood away from the cavernous sinus. This is a potential efferent control mechanism by which SBC can be implemented, fine–tuned or even suppressed by switching blood flow between the two routes. The question is what afferent signals are involved? Independent manipulations of brain and trunk temperatures in goats showed that **Tbrain** provides the exclusive **thermal input** into the control circuit of SBC: trunk temperature signals had no effect [297]. Thus, in the hyperthermic range, the brain has the potential to offset its own temperature from that of the rest of the body core.

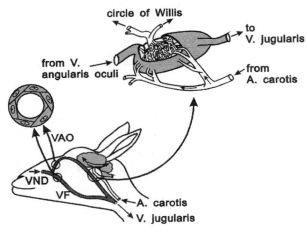

Fig. 8.2. The carotid rete heat exchanger in artiodactyls, and the routes of venous return from the upper respiratory tract. The arterial blood destined for the brain passes through a rete of thin–walled arteries, which are embedded in the cavernous sinus. The cool venous blood returning from the nose in the dorsal nasal vein (*VND*) can flow via the angularis oculi vein (*VAO*) to the cavernous sinus, in which case the arterial blood flowing to the brain is precooled. An alternative route is via the facial vein (*VF*) directly back to the jugular vein, in which case the heat exchanger is bypassed. Both the angularis oculi vein and the facial vein have several layers of smooth muscle and are innervated by sympathetic fibres. (After [260], with permission)

However, recent observations in free–ranging antelopes indicate that SBC is also subject to **non–thermal inputs**. The decisive point was that SBC often occurred at relatively low body temperature, but was abandoned when the animals developed severe hyperthermia during a chase [267,339]. Apparently, the angularis oculi sphincter can close and SBC be suppressed in spite of high Tbrain, for example during strenuous exercise. In this case, even a strong thermal drive to SBC is overridden by a general non–thermal activation of the sympathetic system. Thus, SBC appears possible only in the absence of general sympathetic excitation; then it is fine–tuned by Tbrain.

In the years following the discovery of SBC, it was intuitively assumed that the **biological significance of SBC** lies in protecting the allegedly more vulnerable brain from thermal damage [65]. However, this view is difficult to reconcile with the fact that SBC was not employed when the brain was most in danger of being damaged by high temperature, but prevailed in everyday situations of rest or moderate activity. On the one hand, fighting, chasing or being chased are the only conditions in which larger, euhydrated, non–domestic animals in the wild work hard enough to develop substantial hyperthermia. On the other hand, all three conditions are certainly associated with high levels of sympathetic activity, which are incompatible with the development of SBC. The consequence is that, in non–domestic and free–ranging animals in their natural environment, severe exercise hyperthermia and significant SBC appear mutually exclusive.

The apparent dilemma gave rise to another concept that fits SBC in with the general organization of the thermoregulatory system [260]. SBC also affects the temperature of the rostral brain stem, whose temperature sensors provide a significant fraction of the input signals activating heat loss mechanisms. The facilitation of **SBC** during rest and its inhibition during strenuous exercise alter the input to the controlling system, and could be seen as **a means of adjusting the activity of heat loss mechanisms**. This concept is dealt with in more detail in Chap. 12.

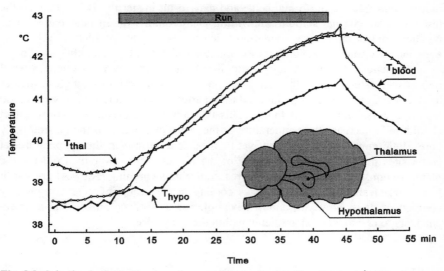

Fig. 8.3. Selective brain cooling in a dog running on a treadmill at 7.2 km·h^{-1}, 20% slope. Air temperature 25 °C, relative humidity 63%. Thalamic temperature (*Tthal*) was considerably higher than hypothalamic temperature (*Thypo*). *Tblood* was measured in a carotid artery, and Tthal and Thypo at the indicated sites (*inset*). (After [12], with permission)

The carotid rete as the essential structure for cooling the **whole brain** is not restricted to artiodactyls but occurs also in felids. Rudiments exist in canids and other fissiped carnivores outside of the cat family [13,103]. Consequently, the degree of whole brain cooling is attenuated in dog [12]. However, there is a second, conductive component. The large surface of the cavernous sinus is in close contact with the base of the brain, and was the likely source of unregulated **local, direct cooling of the rostral brain stem** observed in dog, rat, rabbit, horse and squirrel monkey [15,82,83,137,324]. With the exception of dog, its degree was always small (Fig. 8.3).

An unsettled question is whether SBC occurs in **humans** [61,75]. The problem is that Tbrain cannot be measured directly in healthy subjects and thermal conditions, in which SBC could be expected to occur. As mentioned above, tympanic temperature is a questionable substitute. A priori, the odds appear to be against whole brain cooling: the nose and its mucosal surfaces as the source of cool venous blood are small in relation to the mass of the brain, a specialized heat exchanger does not exist, and a substantial fraction of the blood supply to the brain is via the basilar artery, which has no contact with cool venous blood.

8.2.2
Testes Cooling

The normal intraabdominal temperature of mammals is too high for spermatogenesis; raising testes temperature to core temperature reduces sperm numbers, motility and fraction of normal cells, and may result in sterility [453]. Placing the testes into an external scrotum and fitting the scrotum with an independent temperature control circuit was evolution's first answer to this problem [322]. However, the cetaceans have reinternalized the testes and developed another solution.

In bottlenose dolphin, blood is supplied to the testes by a spermatic arterial plexus that extends from the lumbar aorta and coalesces into a single testicular artery. A venous plexus is juxtaposed to the arterial plexus and returns cool blood from the dorsal fin and tail flukes. This is again a heat exchanger whose effect should be to decrease the temperature of the testes below that of the general body core. Both temperatures could not yet be measured. However, the temperatures at colon sites in the immediate vicinity of the plexus were lower than outside the plexus region, and the differences increased during swimming [386]. At present, it is not yet clear to what extent testes cooling in cetaceans involves a regulatory component. It is, however, another example for the utilization of local heat exchangers to create thermal gradients within the body core.

9 Behavioural Control of Heat Exchange with the Environment

As mentioned before, the thermoneutral zone comprises the central range of environmental conditions in which temperature regulation is achieved entirely by control of dry heat loss. The limits of the range are defined by the lower and upper critical temperatures, below or above which heat production or evaporative heat loss must increase in order to maintain the thermal balance [503].

The thermoneutral zone in resting humans is, by and large, also the zone of thermal comfort [211], and it is little short of commonplace that homeotherms prefer thermal conditions easing the maintenance of normal body temperature [76,428]. Active measures directed to arrive at such conditions are referred to as thermoregulatory behaviour. However, the significance of behaviour extends beyond avoiding thermal discomfort. Autonomic mechanisms like shivering and non–shivering thermogenesis, or panting and sweating, are costly in terms of energy and water, and long–lasting environmental stresses can severely drain the body's stores, if the strain on autonomic mechanisms is not alleviated by appropriate thermal behaviour. In essence, **behaviour is a means of reducing the costs of autonomic thermoregulation**, and it does so by modifying the heat exchange with the environment. The diversity of behavioural adaptations is as large as the number of species and habitats, and the following examples provide just a few guide lines to analyze individual patterns.

9.1 Locomotion

An obvious answer to a harsh environment is to move to a more moderate one. Sheep seek shelter from cold wind behind shrubs, and in the open plains of Africa, small gazelles compete for the shade of single trees during the hot hours of the day. However, the pasture behind the shrubs or under the trees may be worse than in the open field. Yellow baboons sleep on acacia trees, whose sparse canopy does not provide much shelter from the cold morning wind. From a thermoregulatory point of view, the dense shrubs next to the trees would be preferable. However, shrubs are also perfect hiding sites for predators like lions and leopards [478]. In other words, thermal behaviour can incur a price and may even become too expensive, and often animals have to **compromise between conflicting demands** [276].

9.2
Orientation and Posture

The fur of the yellow baboons is dense on the back, and thin on the chest and abdomen. Thus, when air temperature is low and wind speed is high in the early morning, they face away from the wind and sit in a hunched posture: the spine is curved, the chin is dropped against the chest, and the long limbs and tail are held close to the body. The rationale is to reduce the **effective body surface area**: the heat exchange with the environment is a function of the size of the exposed body surface, and by assuming a ball–like posture, heat loss to the environment is minimized. Later in the morning, the baboons face the sun, straighten their backs, raise their heads, and extend their limbs: the radiant heat gain by basking in the sun exceeds the convective heat loss from the enlarged surface area [478].

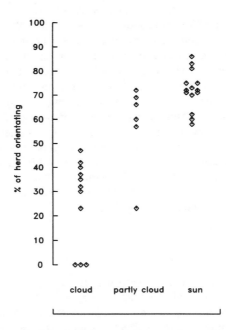

Fig.9.1. Orientation of the long axis of the body towards the sun. On a partly cloudy day, a herd of 150 grazing springbok was observed in 10–min intervals from 9:00 to 15:00 h. When the sun was not obscured, solar radiation was up to 1000 W·m^{-2}, and the percentage of animals orientating the long axis towards the sun was largest. (After [229], with permission)

Springbok, thinly furred gazelles of 20 to 30 kg body mass, occupy regions of intense solar radiation. Orientating the long axis of the body towards the sun reduces the profile area exposed to direct radiation and thus the radiant heat load. Figure 9.1 shows that orientation is part of springbok's behavioural repertoire.

Pictorial descriptions of free–ranging animals like the ones mentioned face the problem that the benefit of appropriate behaviour appears obvious but is difficult to quantify. This is not the case in a laboratory setting. Figure 9.2 compares the results of two studies in rabbit. The upper set of data was collected from animals restrained in a holder, enforcing a spread, open position even in a cold environment [150]. The lower set of data comes from experiments in which the animals could adopt typical and appropriate postures [325]. The difference is striking: heat production of the crouched rabbits in the cold was nearly 50% lower than in the restrained ones, and both groups maintained the same constant body temperature.

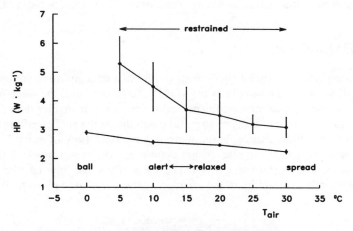

Fig.9.2. Air temperature, body posture and heat production. *Upper curve* Rabbits were restrained in a holder and exposed to air temperatures (T_{air}) from 5 to 30 °C. *Lower curve* Rabbits of the same breed and body mass were placed in a chamber large enough to allow postural adjustments to T_{air} between 0 and 30 °C, resulting in relatively lower heat production (*HP*). (Upper curve after [150], lower curve after [325], with permission)

9.3
Wallowing and Saliva Spreading

Heat–exposed pigs do not sweat, and pant poorly. However, they have the habit of wallowing in mud. It sticks to the sparse and bristly coat and comes in close contact with the skin. Thus, the heat for the evaporation of water contained in the mud is supplied preferentially by the skin. Evaporative cooling rates of 600 W·m^{-2}, persisting for more than an hour, were measured after a single application of mud. Wallowing thus transforms the pig from an animal with limited autonomic capacity into one whose rate of evaporative heat loss matches that of a heavily sweating human. It is a perfect solution for a more sedentary species: the water required for cooling is not taken from the body's water store, and, in contrast to sweating, the electrolyte balance is not affected [244].

Rats and other rodents also do not sweat and have little capacity for panting [153]. During heat stress, saliva is secreted at a high rate and spread on the fur and the vascularized surfaces of the feet. The response is surprisingly efficient: at 40 °C air temperature, intact rats maintained body temperature at 40.5 °C for more than 3 h. However, animals in which the ducts of the salivary glands were ligated, had to be removed from the heat after 1 h because body temperature showed an explosive rise to 41.3 °C [161]. Saliva spreading is of major importance also in marsupials. In a hot environment, big red kangaroos lick the forearms and legs, which have elaborate superficial networks of blood vessels [370]. In bandicoot, a small parameloid marsupial, the evaporative heat loss attributable to salivation accounted for 40% of the total in an environment of 40 °C air temperature [242].

9.4
Social Behaviour

Neonates born in litters regulate body temperature in a cold environment by group huddling [428]. A good example is pig (Fig. 9.3). In the cold, the litter forms a dense pile, whose surface area, and hence heat loss, is much smaller than the sum of the individual surface areas. The stressful positions on the surface of the huddle are occupied on a time–sharing basis: there is a constant flow of animals to and from the warm centre of the pile. In hot conditions, the group breaks into individuals, and the total surface area increases greatly. Thus, the litter has the advantage of considerable variability of the effective surface area for a given total mass.

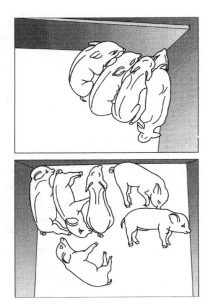

Fig.9.3. Group of young pigs in cool (*top*) and warm (*bottom*) conditions. (After [245])

The **energy saving by huddling** was measured (Fig. 9.4). The lower critical air temperature of a single newborn pig was near 35 °C. On exposure to 20 °C air temperature, its heat production increased from approximately 5 to 8 W·kg⁻¹. However, when the piglet got more and more companions, the heat production per unit of mass declined, and the huddling litter of six at 20 °C air temperature had essentially the same mass–specific heat production as the single piglet at 35 °C [354]. Geometric considerations show that a group size of five or six individuals is sufficient; the additional benefit by increasing the number is small [79].

Such data are suited best to illustrate a salient feature of behavioural regulation. Newborns can respond to cold by shivering or non–shivering thermogenesis. However, this is nearly useless for an individual if it is poorly insulated and unable to conserve heat. The behaviour of a group solves the problem: body temperature is maintained at low cost, and more energy is available for growth. **Body temperature and the cost of maintaining it are controlled by behaviour** [428].

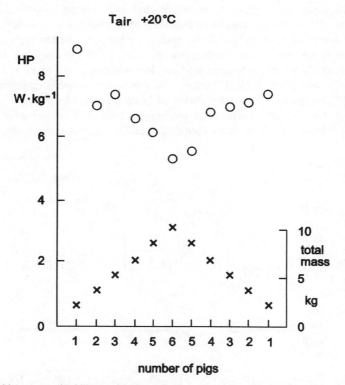

Fig.9.4. Cold exposure, huddling and heat production. Six piglets (body mass 1.2 to 2.6 kg) were placed in a metabolic chamber, one after the other at 30–min intervals. After the complete group had been together for another 30 min, one after the other was removed at the same intervals. *Lower part* Total body mass in the chamber; *upper part* Heat production (*HP*) per unit of body mass decreased with increasing number of animals. (After [354], with permission)

Wild pigs give birth to litters of five or more without the thermal comfort of a shed. They build elaborate nests, small enough to enforce close thermal contact between the sow and her offspring. It is instinctive behaviour: an isolated piglet of age 5 days built its own nest [159]. Nest–building and burrowing are widespread modes of thermoregulatory behaviour in many small species; they give effective protection against cold, heat – and predators.

9.5
Operant Behaviour

An obvious question is whether behavioural responses are exclusively triggered by skin temperature, or whether also **deviations of internal body temperature** are "felt" to guide appropriate behaviour. While the verbal answers of humans are straightforward and clearly positive (Chap. 12), experimenters must know how to question animals properly [428]. A typical setup gives a trained animal the choice to alter the temperature of its environment, and as a consequence, skin temperature. Figure 9.5 shows the results of experiments in sheep that had learned to obtain short bursts of infrared heating by pressing a bar with their muzzles. The animals were shorn, and in the cool environment of 15 °C, pressed the bar nearly once a minute. When brain temperature was lowered from 39.5 to 38.5 °C by infusion of cold saline into a carotid artery, the frequency of bar pressing nearly tripled [17]. The result is clear: deviations of brain temperature aroused sensations motivating the animals to alter their environment. Similar results were obtained from temperature variations of the spinal cord and the medulla oblongata [84,309].

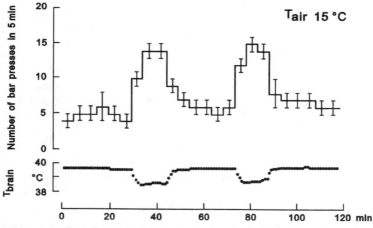

Fig. 9.5. Frequency of bar pressing for infrared heating in a cool environment at normal or low brain temperature (*Tbrain*). Data of 18 experiments in shorn sheep. (After [17], with permission)

10 Autonomic Control of Dry Heat Loss from the Skin

10.1 Pilomotion

As discussed in Chapter 7, fur provides effective protection against external cold. A handicap to the wearer is that the principal features determining its insulating value, spatial density and length of the hairs, require considerable time to adapt to a new thermal environment. The problem is somewhat alleviated by pilomotor activity: the contraction of smooth muscle fibres inserting at the roots of the hairs erects the shafts and can nearly double the fur thickness [41]. The response can occur within seconds upon sudden cold exposure [400]. Its efferent arm operates via sympathetic fibres [158]. However, as mentioned in Chapter 7, doubling the thickness does not double the insulation. Thus, in contrast to birds, the capacity of pilomotion to modify the heat flow to the environment is modest in mammals.

10.2 Peripheral Blood Flow

The basic principle is simple. The rate of dry heat loss by conduction, convection and radiation is always influenced by the temperature gradient between the skin and the environment. Thus, heat loss in the cold can be reduced by lowering skin temperature (T_{skin}), and conversely in the warmth, maintaining the skin warmer than the environment permits dry heat loss of at least some magnitude so that less water is required for evaporative heat dissipation. T_{skin} is of course a function of skin blood flow. However, it is not the only determinant [502].

10.2.1 Skin Blood Flow and Skin Temperature

We consider the steady state of a non–sweating skin region at the limbs or appendages like the tail or the ears. The external heat flow from the skin to the air equals the internal heat flow from the blood to the skin:

$$\text{external heat flow } H_e = \text{internal heat flow } H_i \text{ [W·m}^{-2}] \qquad (10.1)$$

The **external heat flow** is the product of two factors: the external heat flow coefficient h_e (lumping together the natural and forced convective heat loss coefficients of Chap. 6), and the difference between T_{skin} and air temperature (T_{air}):

$$H_e = h_e \text{ [W·m}^{-2}\cdot°C^{-1}] \cdot (T_{skin} - T_{air} \text{ [°C])} \qquad (10.2)$$

The **internal heat flow** is the product of two other factors: the internal heat flow coefficient h_i (similar to, but not identical with, the internal conductance defined in Chap. 7) which depends on skin blood flow, and the temperature difference between the arterial blood arriving at the skin (T_{art}), and T_{skin}:

$$H_i = h_i [W \cdot m^{-2} \cdot {}^\circ C^{-1}] \cdot (T_{art} - T_{skin} [{}^\circ C]) \qquad (10.3)$$

Combining Eqs. 10.1, 10.2 and 10.3 yields:

$$T_{skin} = \frac{(h_e \cdot T_{air}) + (h_i \cdot T_{art})}{h_e + h_i} \qquad (10.4)$$

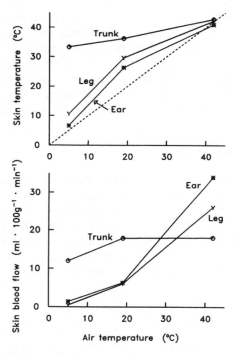

Fig. 10.1. Skin temperature (*top*) and skin blood flow (*bottom*) of conscious sheep in cold, neutral and hot environments. The *interrupted line in the top panel* indicates identity of air and skin temperatures. Blood flow was measured by labelled microspheres. (Data from [162,166])

Of the four factors involved, T_{air} and h_e are subject, at least to some extent, to behavioural control. In a given environment and at constant T_{art}, however, variations of T_{skin}, and hence the control of heat loss to the environment, are mainly accomplished by changes in the internal heat flow coefficient h_i, that is, by **changes in skin blood flow**. Figure 10.1 gives an example.

Fig. 10.1 shows that the environment–induced changes in T_{skin} and skin blood flow (SBF) were small under the thick coat of the trunk, and large at the sparsely furred legs and ears. In the cold, SBF in the legs and ears was close to nil, but multiplied by a factor of 30 or more in the heat. In all mammals, the range of regulatory variations of SBF is particularly wide in those regions of the body, which have large ratios of surface area to mass and volume (Fig. 10.2). Fingers and appendages like the ears, or the flippers of the seal, or the tail of the rat are **thermal windows**, whose potential rate of heat dissipation to the environment are determined by local SBF, in accordance with the thermal state of the body.

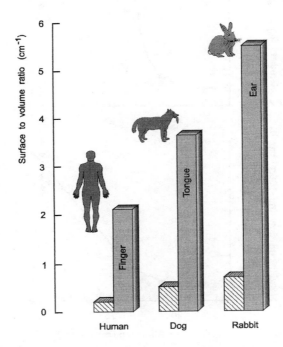

Fig. 10.2. Ratios of surface area to volume of the whole body (*cross–hatched bars*), and single body regions involved in fine–tuning heat loss in different species. (Data from [214])

In a hot environment, however, Eqs. 10.2 and 10.3 set the limits to physiological control of dry heat loss. On the one hand, the external heat flow from the skin to the environment requires T_{skin} to be higher than T_{air}. On the other hand, the internal heat flow from the body core to the skin requires core temperature to be even higher than T_{skin}. Thus it is the **upper tolerance level of core temperature**, which finally determines whether an animal can maintain a sizeable dry heat loss in a hot environment [490]. This was certainly not the case in some experiments of Fig. 10.1: at 40 °C T_{air}, the gradient between T_{skin} and T_{air} was too small to establish a sufficient rate of dry heat loss, in spite of large SBF. Thus, the closer T_{air} comes to T_{core}, the more evaporative heat loss mechanisms must step in to dissipate heat, either by cooling the skin (sweating) or by panting. The adjustment of **dry heat loss** by SBF–induced modulations of T_{skin} is of physiological significance mainly **in cold or thermoneutral** environments (Fig. 10.3).

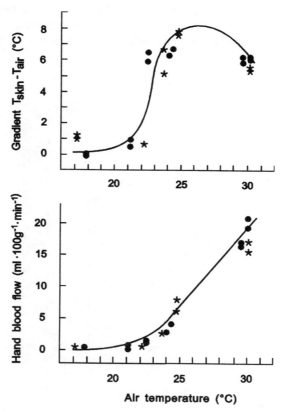

Fig 10.3. Hand blood flow (*bottom*) and the gradient between skin and air temperature ($T_{skin} - T_{air}$, *top*) plotted vs. air temperature in two human subjects. Hand blood flow was measured by plethysmography, and its changes essentially reflect changes in skin blood flow. (After [124], with permission)

Figure 10.3 shows that small increases in SBF in the 21–25 °C range of T_{air} caused large increases in the gradient between T_{skin} and T_{air} which, in the particular experimental setting, grossly reflected the rates of dry heat loss. In the 25 to 30 °C range of T_{air}, however, the gradient (and heat loss) actually decreased, in spite of the further large increase in SBF.

10.2.2
Anatomy of Peripheral Circulation

The cutaneous microvasculature features particular elements to accommodate the required large changes in SBF. **Arteriovenous anastomoses (AVAs)** are precapillary structures connecting the arterial and venous sides of circulation. They allow arterial blood to enter the venous part of the circuit without passing through capillary beds. AVAs are present in many tissues and organs, but are most numerous in the skin, and **abundant in acral regions** like ears, nose, fingers, paws or the flippers of aquatic mammals, that is, in the thermal windows. They can be patent or closed (Fig. 10.5): AVAs are richly innervated and have not only circular smooth muscle but also an inner longitudinal muscle layer. While the diameter of capillaries is 10 µm or less, patent AVAs have diameters of up to 150 µm, and provide a low–resistance pathway for high rates of blood flow [164,348]. In a panting dog, 70% of total tongue blood flow passed through AVAs [394].

AVAs occupy the deeper layers of the skin and must divert part of the blood flow from the capillaries and the skin surface where the heat exchange with the environment is to take place. However, this is not as counterproductive as it seems: AVAs drain into superficial veins, and it is there where the blood gives off part of its heat to the surface. Thus, patent AVAs serve to increase blood flow and heat loss in downstream superficial veins.

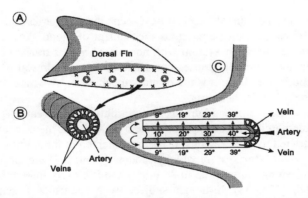

Fig. 10.4 A, B, C. Arrangement of blood vessels in the flipper of dolphin. In the cold, central arteries and adjacent trabeculate veins form heat exchangers. During heat loads, the exchangers are bypassed because the blood returns via superficial veins, indicated by *crosses*. (After [449], with permission)

Another feature of the vascular system in the limbs and appendages is that it provides **two routes of venous return** to the trunk. One route goes via deep and thin–walled veins which, for example in the flippers of the dolphin, wreath around the central arteries [449]. In a cold environment, blood flow and velocity in veins and arteries are low, and the system acts as a **countercurrent heat exchanger**: heat is transferred from the arterial to the venous blood and returns back to the body core

(Fig. 10.4). This can be seen in the light of Eq. 10.3: because the arterial blood in distant parts of the legs or appendages assumes nearly the same temperature as the skin, the internal heat flow is reduced to a minimum. The decrease in arterial temperature evolves along the length of the flipper so that not just the skin but also muscles cool down. The flipper as a whole becomes part of the body shell. Countercurrent heat exchange between central arteries and concomitant veins occurs also in the limbs of terrestrial animals and was, in fact, first seen in humans: along the brachial artery, blood temperature decreased as much as 3 $°C·10$ cm^{-1} [25].

During **heat stress**, the rate of blood flow in the arteries to the skin increases, and the now thicker arteries in the flipper may be thought to compress the trabeculate veins so that the **venous return is diverted to superficial veins**. In terrestrial mammals, the dilation of superficial veins is caused by sympathetic relaxation and direct temperature effects on the vessels.

The heat–induced larger acral blood flow through AVAs and the diversion of the venous return to superficial veins must be seen in context: the large flow raises T_{skin} not just at the distal parts, but at the full length of the limbs. The low blood velocity in dilated veins prolongs the time for heat exchange between the blood and the skin. In hyperthermic humans, more heat was lost from the arm than from the hand, despite the AVAs being most abundant in the hand [222,528].

10.2.3
The Central Drive

All vessels of the peripheral vasculature except the capillaries are supplied by sympathetic fibres. A nearly ubiquitous mechanism of blood flow control is the **adrenergic vasoconstrictor system** with norepinephrine as transmitter [274]. Increasing sympathetic activity causes constriction, and withdrawal of tone is followed by vasodilation [170]. It is the primary mechanism by which SBF is modulated in the control of dry heat loss.

In the skin of heat–stressed humans and possibly other sweating primates, the role of the adrenergic vasoconstrictor system as the sole neurogenic means of control is restricted to acral regions. It is supplemented at the limbs and the trunk by an **active vasodilator system** which accounts for most of the large increase in total SBF in sweating humans (up to 8 $l·min^{-1}$) and is closely linked to the activity of sweat glands [274,434]. In human subjects with congenital absence of sweat glands, SBF during hyperthermia increased much less than in normal subjects [62]. Present evidence suggests that active cutaneous vasodilation is mediated by cholinergic fibres and involves the local release of nitrous oxide from the endothelium [283,284]. Active vasodilation as a response to whole body heating was also observed in the rabbit ear [122].

10.2.4
Local Effects of Skin Temperature

Figure 10.5 shows results of a study in which the velocity of blood in finger arteries was recorded as an index of SBF. With the subject in a thermoneutral environment and the hands at relatively high temperatures (Fig. 10.5, top), the velocities in both fingers showed synchronized large fluctuations. Their synchrony points to bursts of efferent sympathetic activity as the common cause, and their magnitude suggests AVAs as the targets. The fluctuations themselves are characteristic of a thermoneutral situation: the AVAs were permanently closed during cold exposure and permanently patent in the heat, but alternated between both states in thermoneutrality, at a frequency of two to three cycles·min^{-1} [31].

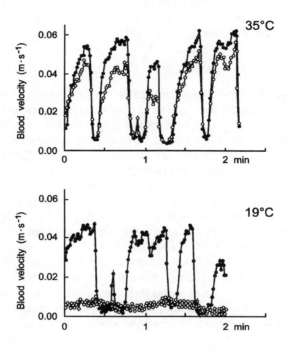

Fig. 10.5. Local temperature effect on finger blood flow of a human subject. Simultaneous recordings of blood velocity (indicative of blood flow) in 3rd finger radial arteries of both hands. A *single symbol* shows average velocity during one heart cycle. *Top* Left hand (*filled circles*) exposed to 24–27 °C air temperature, right hand (*open circles*) in water at 35 °C; *bottom* Left hand exposed to 24–27 °C air temperature, right hand in water at 19 °C. (After [32], with permission)

However, SBF is not entirely under neural control but also subject to direct effects of temperature on the blood vessels. Putting the right hand into water at 19 °C abolished the fluctuations, and the blood velocity in the cooled finger decreased to a minimum, most likely determined by the nutritive flow to capillaries at permanent closure of AVAs (Fig. 10.5, bottom). The mechanisms mediating local temperature effects on skin blood flow are likely to be different in acral and non–acral regions, and not necessarily the same during intense cooling and warming [284,391]. However, the local temperature effect disappears when high T_{core}

abolishes the vasoconstrictor tone in the skin vessels and SBF is near its maximum [413,498]. This points to **sympathetic neurotransmission** as the primary target of local temperature effects. In general, small superficial arteries and veins have a high density of $\alpha 2$–adrenoceptors, while $\alpha 1$–adrenoceptors prevail in deep vessels. Both types of receptors have a similar affinity to norepinephrine, but are differently sensitive to temperature. Local cooling enhances contractions induced by $\alpha 2$–adrenoceptor stimulation, while it inhibits the $\alpha 1$–adrenoceptor response [129]. The effect is possibly also a key factor in the diversion of the venous return from superficial veins in the heat to deep veins in the cold: local cooling constricts superficial vessels, and dilates deep veins [512].

10.2.5
Specialized Surface Regions

At very low T_{air}, the requirements dictated by the heat balance of the whole body must compromise with the necessity to prevent acral tissues from **freezing** [216]. In arctic dog exposed to $-30°C$ T_{air}, local T_{skin} at the face and the legs was between 1 and 10°C [251]. Thus SBF in bare or thinly furred parts of the body is regulated at levels just sufficient to maintain T_{skin} above 0 °C. Anecdotal observations in arctic birds suggest that this high degree of regional adaptation is lost unless the birds are permanently exposed to the cold environment [249].

A possibly related phenomenon is observed in inhabitants of moderate climates during exposure to intense cold. When T_{skin} at a finger or ear falls below 15 °C, intense vasoconstriction is periodically (≈ 20–min) interrupted by brief (≈ 1–min) episodes of vasodilation. This so–called **Lewis reaction** is assumed to protect the skin from cold injury. The dilation occurs mainly in regions rich in AVAs and could be caused by a direct relaxing effect of cold on the AVAs' smooth muscle [33]. The sudden flush of warm blood then reestablishes vasoconstrictor control [512]. In the tongue of dog, a dilatory effect on AVAs of cooling occurred in the 28 to 40 °C range of surface temperature which was not nervously mediated [294,394]. It could contribute to sustaining high blood flow in the protruding tongue when a dog is running in a cold environment.

It is not quite clear what happens to SBF when, in a very hot environment, T_{skin} exceeds T_{core}. Desert jack rabbits have extremely large ears (25% of total body surface), and observations reported local vasodilation in the shade, but constriction in the sun, which would help to reduce the heat flow into the body [441]. Laboratory experiments in the tail of rat showed a similar pattern [399]. However, results in sheep were at variance [171], and also in humans, no persistent reductions of SBF were observed when T_{skin} exceeded T_{core} [31,364].

11 Autonomic Control of Evaporative Heat Loss

In a hot environment or during intense exercise, mammals must resort to evaporative mechanisms of heat dissipation to maintain thermal balance. As detailed in Chapter 6, the cooling power of evaporation is large, and except in a hot and humid environment, all terrestrial mammals can balance at least resting heat production for a limited period of time entirely by evaporative heat dissipation, and some species do much better.

Evaporation occurs on the mucosal surfaces of the **respiratory tract**, and on the **skin**. Both channels have a passive component (perspiratio insensibilis): even in a cold environment, the exhaled air is saturated with water vapour, and water passes through the skin by diffusion. During heat stress, one or both have an active, controlled component: panting and sweating. The extent to which the channels are used, varies greatly. There is a **phylogenetic aspect**: primates and equids rely primarily on sweating, while carnivores and rodents pant [416]. Another clue is **body mass**: smaller bovid species primarily utilize panting, while sweating is quantitatively more important in larger ones [418]. The evaporative repertoire of marsupials includes saliva spreading: red kangaroos pant and spread saliva during rest in a hot environment, but sweat during exercise [114].

11.1
Sweating

The capacity for sweating is large in **primates** and **equids**. Maximum sweat rates (SR) of 1 and 3 $l \cdot m^{-2} \cdot h^{-1}$ have been reported in humans [42] and horse [226], respectively, whose complete evaporation would dissipate nearly 15 times resting heat production. In contrast to others, primates and horses can maintain high work rates also in a warm environment, and sweating is certainly the decisive prerequisite: in humans habitually active in a temperate environment, maximum metabolic rate and maximum SR were closed coupled [104]. However, the rate of evaporative cooling depends not only on SR but also on the water vapour pressure and velocity of the surrounding air (Chap. 6). A natural environment permits rarely the evaporation of 3 $l \cdot m^{-2} \cdot h^{-1}$, which means that part of the sweat drips off and is wasted. Thus, in horse, the capacity for sweating is somewhat oversized in relation to the usually prevailing cooling power of the environment.

This is the case at the other end of the mass range. Patas monkey is a small primate of approximately 4 kg mass, living in the open African savanna. It is reputed to be active all day and the fastest running non–human primate. Maximum

SR is just 0.3 $1 \cdot m^{-2} \cdot h^{-1}$, but the ratio of surface area to body mass is nearly ten times larger than in horse. Figure 11.1 shows that a cutaneous evaporative heat loss of nearly 40 $W \cdot kg^{-1}$ enabled the monkey to maintain a high speed even in a 55 °C environment, when convective and radiant heat gains from the environment added considerably to the metabolic heat load. The increase in core temperature (Tcore) was not greater than 2 °C [320].

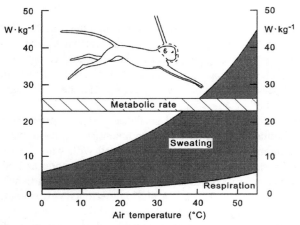

Fig. 11.1. Metabolic rate and evaporative heat loss by sweating and respiration of a patas monkey, running on a treadmill (speed 11–13 $km \cdot h^{-1}$) at air temperatures from 0 to 55 °C and low water vapour pressure. When air temperature exceeded body temperature and heat was gained from the environment by radiation and convection, the rate of evaporative heat loss was larger than the metabolic rate. Based on approximately 30 runs of 11–22 min duration. (After [320], with permission)

In the order **Artiodactyla**, members of the family Suidae (including domestic pig) show no thermoregulatory sweating. In the families Camelidae and Bovidae, however, sweating can be an important route of heat loss. SRs of 0.2 $1 \cdot m^{-2} \cdot h^{-1}$ were found in euhydrated resting camel [445], and the order of magnitude is the same in large bovids [418]. Also some smaller species like sheep and goat, if adapted to hot climates, have SRs which are significant in view of the more sedentary way of life [118,232].

11.1.1
Types of Sweat Glands and Modes of Transmission

Ekkrine sweat glands open **directly to the skin** and occur at high density over most of the body surface in primates, and at glabrous areas such as the foot pads in non–primates. The gland consists of a secretary coil and a duct. The coil produces **precursor sweat**, containing sodium and chloride at plasma concentrations. In the proximal part of the duct, the sodium concentration is reduced by an active pump, and chloride follows passively so that the fluid becomes hypotonic [398,433]. If the skin is fully wetted, the superficial end of the duct can be partially obstructed by skin swelling; the phenomenon is called **hydromeiosis** and is likely to be responsible for the decline in SR during prolonged exposure to a hot and humid environment [80,363,534].

Ekkrine sweat glands are diffusely innervated by sympathetic fibres. The stringent nervous control is reflected in the synchronization of sweat expulsions over the general body surface, following bursts of activity in sudomotor fibres and becoming visible at low SR (Fig.11.2).

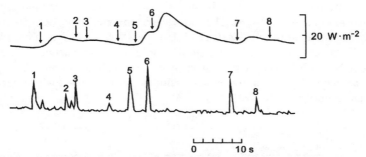

Fig.11.2. Sympathetic activity in a nerve supplying the foot (*bottom*), and rate of evaporation on the dorsal foot (*top*), in a human subject at 28°C air temperature. Bursts of sympathetic activity generated expulsions of sweat after a delay of 2 to 3 s. A single expulsion caused increasing evaporation, followed by slow decline when the nerve was silent. (After [384], with permission)

The degree of sympathetic activity reflects the **central efferent drive** to sweating, which originates in the rostral brain stem. At higher levels of heat stress, the bursts fuse and a continuous flow replaces the discrete sweat expulsions. Although **acetylcholine** serves as the principal transmitter, evidence exists of a supportive role of adrenaline. In stump–tailed macaque, an anthropoid primate, SR during exercise was 50% larger than the maximum during exposure to external heat. The increment was lost after denervation of the adrenals, and restored by intravenous infusion of adrenaline [419].

Apokrine sweat glands are associated with **hairs** and open into common pilo–sebacious orifices. In primates they are restricted to some special regions such as the axillae. In furred non–primates, however, they are present over most of the body surface and the sole source of thermal sweat in species utilizing this avenue of evaporative heat loss. Apokrine glands are endowed more than ekkrine glands with myoepithelial cells whose rhythmic contraction causes discrete discharges of sweat at intervals of several minutes. This type of sweating is seen most clearly in goat, sheep and other small bovids whose capacity for sweating is limited [256,418].

Apokrine sweat glands are under **adrenergic** control. In horse and donkey, the glands are innervated by sympathetic fibres. The more potent transmitter, however, is adrenaline acting on ß–receptors. Thus, it appears that the sweat glands in equids are under dual control by circulating adrenaline and locally released nora-drenaline [37,417]. In contrast, bovine sweat glands have α–receptors, and the adrenal medulla is not involved in mediating the sweating response to heat and exercise. However, the glands have no direct nerve supply. The transmitter is released at some distance from the gland which may explain the somewhat coarse control of sweating in bovids [416].

11.1.2
Sweat Rate: Central Drive and Local Effects

The relationship between afferent thermal signals and SR has been thoroughly investigated in humans [259,434]. The influence of **internal body temperature** exceeds that of **mean skin temperature** (T_{skin}) by a factor of 10 or more [42,274], and a 2 °C increase in T_{core} above the threshold is usually sufficient to generate maximum SR [104]. The central drive may also include a short–lasting non–thermal component [385]: in dynamic conditions, SR responded within seconds to variations in work rate [26,292].

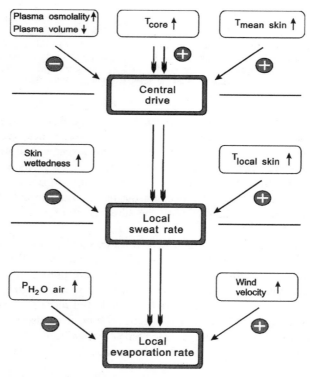

Fig. 11.3. Evaporation on the skin. General factors determining the central drive to sweating, local factors influencing the transformation of the central drive into sweat rate, and environmental factors affecting the rate of evaporation

However, the SR at any site does not depend entirely on the central drive but increases with **local skin temperature**. The effect is large and could be caused by an increased transmitter release per unit of sympathetic activity or a larger response of the gland per unit of transmitter [384]: a 5 °C increase in local temperature nearly doubled the local sweating response to a given combination of T_{core} and mean T_{skin} [358]. Another peripheral factor is the wettedness of the skin (hydromeiosis).

Comparable studies in other species are rare. It seems likely, however, that also in horse and other species capable of maintaining high work loads, the decisive thermal signals arise from core sensors of temperature. In contrast, SRs in desert–adapted and more sedentary species exposed to high solar radiation often correlated with T_{skin} but not with T_{core} [118,127]. Longer–lasting sweating causes **dehydration** if the fluid loss is not compensated by drinking. Relatively small degrees of dehydration reduce plasma volume and increase plasma osmolality, both of which **inhibit sweating**. Chapter 15 deals with the effects of dehydration in more detail.

11.2
Panting

We start with a **resting animal in a thermoneutral or cold environment**. Inevitably, respiration involves the loss of water and heat, because the water content of exhaled air is higher than that of inhaled air. However, the passive loss of water and heat can be minimized, and this strategy is amply used by a variety of species, from small rodents to giraffe [302,442]. Resting animals **inhale and exhale through the nose**. On inhalation, the air is humidified and warmed to T_{core} by passing the mucosal surfaces of the nose. However, the same process simultaneously cools the mucosa. On exhalation, the air leaving the lungs is saturated with water at body temperature and passes the cooler surfaces so that a substantial fraction of its water content condenses. Thus on exhalation, **the nose recuperates** some of the water and heat added to the air on inhalation. It is clear that the process works best in species whose long nose provides small transverse and large axial dimensions of the airways. Furthermore, local blood flow must be relatively low to avoid rewarming of the mucosa before exhalation occurs (Fig. 11.4,0).

After a while in a **warm environment**, increasing respiratory evaporative heat loss (REHL) becomes, in non–sweating species, a necessary means of compensating for the decline in dry heat loss. The first step is **to secrete more water** and **to increase mucosal blood flow** [162]. The nasal mucosa contains a complex system of arterial and venous retia and arteriovenous anastomoses [110], serving to adjust mucosa surface temperature to the need for heat dissipation [271]. The effect of higher blood flow is that the mucosa warms quickly after inhalation, and consequently recuperates less heat and water on exhalation. Thus, more heat is dissipated, in spite of **inhalation and exhalation** still occurring entirely **through the nose** (Fig. 11.4,1).

IN   EX

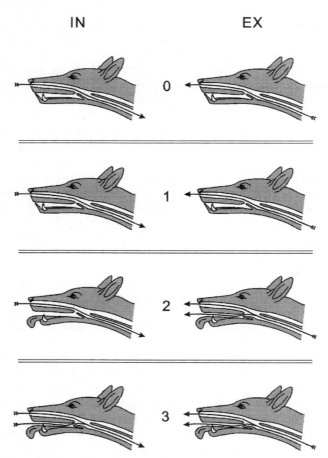

Fig. 11.4. Patterns of air flow through nose and mouth on inhalation (*left*) and exhalation (*right*) in dog. **0** Resting in a thermoneutral or cold environment; **1** Resting in a warm environment; **2** Resting in a hot environment; **3** Exercising in any environment. (Modified after [491])

In a **hot environment**, even a resting animal must do more, and the appropriate way is to increase ventilation, because REHL is, of course, a function of the rate of air flowing over the evaporating surfaces. At least in dog, inhalation is still entirely through the nose, but **exhalation** takes place increasingly **through the open mouth** [149] so that the resistance to flow becomes smaller and the process of recuperation in the nose is bypassed (Fig. 11.4,2). This is the characteristic pattern of **rapid shallow panting**. In submaximal heat stress, high ventilation is accomplished by greatly increased respiratory frequency combined with reduced tidal volume, so that alveolar ventilation and acid–base balance of the blood remain essentially unchanged. In a very hot environment and at high Tcore, however, total ventilation increases further, and includes larger alveolar ventilation [167].

A potentially harmful side effect of **second–phase panting** in a resting animal is **respiratory alkalosis**: arterial pCO_2 can decrease from its normal value of 5.3 kPa to less than 3 kPa, and arterial pH increase from 7.4 to 7.6 and more (Fig. 11.5). At this stage, the feedback between blood gas pressures and alveolar ventilation is abolished [163]. It should be noted, however, that the occurrence of severe respiratory alkalosis is restricted to extremely adverse conditions: in a natural environment, resting animals do not often develop sufficiently high levels of T_{core}. A different situation exists during exercise in a warm environment. Because the ventilatory loss of CO_2 is partially replaced by increased supply, respiratory alkalosis is less of a problem and may even be compensated by metabolic acidosis, adding a chemical to the thermal drive for ventilation.

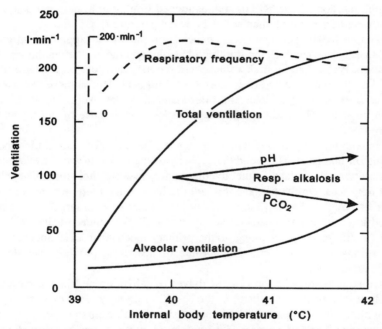

Fig. 11.5. Total ventilation and alveolar ventilation plotted vs. internal body temperature in resting calves exposed to severe external heat stress. At internal body temperature below 40 °C, rapid shallow panting (*inset*) led to increased dead space ventilation (represented by the difference between total and alveolar ventilation). When body temperature exceeded 40 °C, second–phase panting developed: the further increase in ventilation included augmented alveolar ventilation and caused respiratory alkalosis. (After data from [167], with permission)

During exercise, inhalation occurs through nose and mouth, and exhalation mainly through the mouth (Fig. 11.4,3). Humidification and warming of the air inhaled through the mouth are accomplished by large saliva production [455] and concomitant increases in blood flow of the tongue and the buccal surfaces, mainly through arteriovenous anastomoses [394]. Thus the protruded wet tongue of a running dog does not just dissipate heat directly to the environment, but also humidifies and warms the inhaled air [149].

The rate of REHL during exercise consists of two components [120]. The first is obligatory and determined by metabolic rate and alveolar ventilation. The second is facultative and reflects the thermal state of the body and the central drive to increase REHL by augmented ventilation, salivation and mucosal blood flow. In the first part of the experiment shown in Fig. 11.6, REHL increased in parallel with increasing T_{core}, but decreased to its obligatory minimum determined by exercise–related alveolar ventilation, when T_{core} was lowered during continuing exercise. This implies that the rate of ventilation during exercise is not entirely determined by the metabolic rate: in goats exercising at about 50% of maximum metabolic rate for 1 hour, REHL and, as judged from arterial pO_2 and pCO_2, ventilation were larger at high than at low body temperature [123].

Maximum REHL during exercise depends on the water vapour pressure of inspired air (Fig. 6.5) and, as the limiting physiological factor, the maximum rate of ventilation. Exceptionally high values were reported for dog: on a treadmill, 7–8 $l \cdot min^{-1} \cdot kg^{-1}$ were maintained over more than 2 h [541]. This is more than twice as much as in human top athletes and thoroughbred horses exercising at maximum oxygen uptake [121], and explains the remarkable endurance of dogs running in cool environments.

At maximum rate of ventilation in horse, the complete humidification and warming of inhaled air is beyond the capacity of the upper airways. Consequently, full saturation takes place in the alveolar space, making **the lungs** themselves **a site of heat loss**. On the passage through the lungs, blood temperature was found to decrease by 0.4–0.6 °C [226,313]. The decrement times a large cardiac output resulted in a pulmonary heat loss amounting to 20% or more of the total. Thus, even in a species using sweating as the primary mechanism of evaporative cooling, respiration can be a significant route of heat loss, provided ventilation is maintained at a high level.

There is no doubt that the **central drives** to salivation and increased mucosal blood flow and panting originate in the rostral brain stem. However, the conversion of the drive for panting into motor signals to respiratory muscles at the bulbo–pontine levels of respiratory control has remained largely enigmatic [406]. The problem is that two competing demands – metabolic for blood gas control by alveolar ventilation, and thermal for heat loss by dead space ventilation – require independent adjustments of tidal volume and respiratory frequency. While rapid shallow panting can match both demands at moderate levels of heat stress, second–phase panting does not: alveolar ventilation is uncoupled from arterial pCO_2 and metabolic rate [163]. The effect of dehydration on REHL in panting species is similar to that on sweating: the thresholds of T_{core} or T_{skin}, at which panting ensues, are shifted to higher levels (Chap. 15). Interestingly, in species which sweat and pant, the inhibiting effects of dehydration are strong on sweating, and weaker on panting [14,116,261,490].

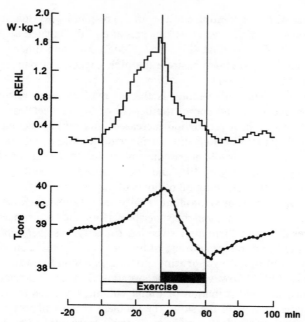

Fig. 11.6. Internal temperature (T_{core}) and respiratory evaporative heat loss (*REHL*) of a goat, walking on a treadmill at 3 km·h^{-1} and 10% slope (*white bar*) at 30 °C air temperature. From min 0 to 35, T_{core} rose because of high heat production, and REHL increased in parallel. From min 35 to 60 (*black bar*), T_{core} was lowered by heat exchangers acting on blood temperature. REHL decreased in spite of continuing exercise to the obligatory minimum determined by alveolar ventilation (min 50 to 60). (After [269], with permission)

11.3
Sweating vs. Panting

If the capacity for heat dissipation is compared in terms of W·kg^{-1}, sweating is, of course, the superior mechanism, because the skin provides a relatively larger surface area for evaporation [491]. In a hot environment, however, a given rate of evaporative heat loss by sweating requires more water than by panting. The reason is that a sweating animal has a cooler skin than a panting one, resulting in relatively smaller dry heat loss. The difference must be made up by additional sweating. In a very hot environment, the cool skin of a sweating animal actually gains heat by convection while the skin of a panting animal may still be warmer than the air [496]. Thus, **panting** is the more appropriate solution to cope with environmental heat loads in regions of **restricted water supply**, in particular if a thick fur helps to limit the effective radiant heat load. However, species indigenous to hot regions and relying entirely on panting must adopt a predominantly **sedentary life**, and restrict higher levels of activity to short bouts whose extra heat production can be temporarily stored [497]. At low air temperature, the disadvantage of the relatively smaller evaporating surfaces of the respiratory system is less significant, partly because the low water vapour pressure of the inhaled air increases the heat loss per unit of ventilation (Fig. 6.5). Sledge dogs and other heavily furred animals like reindeer [273] demonstrate the high capacity of panting species for exercise in a cold environment.

A potential problem with panting is that it involves more respiratory work: higher heat production could curtail the benefit of higher heat loss. However, at least in resting cattle, heat production increased only during second–phase panting, while rapid shallow panting did not incur noticeable metabolic costs [168]. Dogs tend to pant at the resonance frequency of the thorax, at which the energy cost of panting is small [98,149], and the proportional control of REHL is achieved by modulating the duration of bouts of panting [41].

If **exercise in a warm environment** is to be **maintained** for longer periods of time, **sweating** is a necessity, and the superior performance of primates and equids in the heat results from their high sweating capacity. As mentioned above for horse, respiration is an important route of heat loss also in these species: sweating simply adds an **additional evaporating surface**, enabling them to sustain higher rates of exercise [496]. However, intense sweating is always associated with high skin blood flow and redistribution of blood volume from central regions to the compliant venous vessels of the skin. The problem is most important for humans in upright position, and is dealt with in more detail in Chapter 15.

12 Interaction of Various Body Temperatures in Control of Thermoregulatory Responses

Even animals living in a temperate climate are regularly subject to divergent combinations of skin and core temperatures. Because both variables influence all thermoregulatory responses, their modes of interaction are of major practical importance. Another more theoretically orientated purpose of this chapter is to provide data for the next one, which is concerned with the central nervous component in short–term temperature regulation. Input–output data alone cannot reveal how the process of regulation is actually performed in the central nervous system. However, they can set the frame for the neurophysiological approach and pose questions, which await answers from more specific methods [259,466].

12.1
Core Temperature as a Compound Input

In the preceding chapters, the temperature of the body core (T_{core}) was often treated as one input, in spite of the fact that it is actually a compound of inputs generated at spatially separate sites. The reason was that, with two exceptions, the naturally occurring temperature differences between internal thermosensitive sites are likely to be too small to have regulatory significance. The exceptions are selective brain cooling and natural spinal cord warming, which are covered later in subsections of this chapter.

For the start, however, the focus is on experimentally induced large temperature differences in the body core. They have nothing to do with real–life conditions and serve exclusively to analyze the relationship between T_{core} and its input to the controlling system. Figure 12.1 shows heat production (HP) and respiratory evaporative heat loss (REHL) of goats as functions of head and trunk temperatures (T_{head}, T_{trunk}), which were altered independently of each other by heat exchangers incorporated in the blood stream [263]. Both regions included skin thermoreceptors. However, all available evidence indicates that the dominant thermal inputs were generated in the body core: the input attributable to T_{head} by temperature sensors located in the brain stem, and the input attributable to T_{trunk} by spinal cord sensors and other thermosensitive elements in the trunk.

Figure 12.1 shows contour lines to visualize the relationship between the two independent variables (T_{head}, T_{trunk}) and one dependent (HP or REHL). Contour plots are somewhat difficult to read. The principle is that one contour line joins all combinations of T_{head} and T_{trunk} which result in one level of HP or REHL, respectively. Imagine for a moment that the contour lines were straight and vertical:

this would imply that HP and REHL were determined entirely by Ttrunk, while Thead had no influence at all. Conversely, a strictly horizontal course would assign all effects to Thead, and none to Ttrunk. If Thead and Ttrunk contributed equally and linearly, the lines were straight and rising at a slope of 45°.

However, the lines in Fig. 12.1 are curved, running nearly parallel with the trunk axis at high Thead, and nearly parallel with the head axis at high Ttrunk. This implies that HP and REHL were **more influenced by high temperatures** of head or trunk, and less by low temperatures. Thus, in the case of a region attaining a high temperature, all effector outputs are increasingly dominated by this single temperature, while the cooler rest of the body exerts a correspondingly weaker influence. What applies to its components is valid also for **the compound core temperature signal**: its magnitude, per unit change in temperature, **increases exponentially with increasing temperature**.

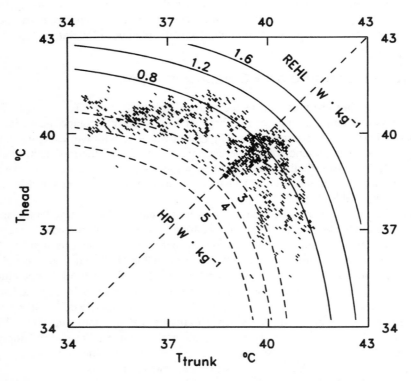

Fig. 12.1. Heat production (*HP*) and respiratory evaporative heat loss (*REHL*) plotted vs. head and trunk temperatures. Head temperature was manipulated by carotid heat exchangers, and trunk temperature by heat exchangers in trunk veins. *Single symbols* represent combinations at which HP and REHL adopted resting levels. The *contour lines* for higher levels of HP and REHL were computed by non–linear regressions. Based on 100 h of data collection in three goats. (After [263], with permission)

The single symbols between the contour lines in Fig. 12.1 represent subthreshold temperature combinations at which the animals neither shivered nor panted. Many symbols are clustered around the line of identity between 39 and 40 °C, which is close to the normal body temperature of goat in a thermoneutral environment. However, the symbols occur also at extreme combinations of Thead and Ttrunk: Thead in excess of 40 °C did not generate panting when it was combined with low Ttrunk, and vice versa. Thus within a certain range, **opposing temperatures** of head and trunk **balanced each other** so that neither HP nor REHL were activated. As a whole, the distribution of the directly determined subthreshold combinations mirrors the shape of the computed contour lines.

However, the significance of balancing temperature combinations extends beyond mere confirmation of the contour lines. If the activities of **two different effector mechanisms**, namely HP and REHL, are checked at basal levels by the same counteracting temperature signals from two sites, then the system behaves as if both HP and REHL follow **one regulated variable**: a compound core temperature signal, processed from signals originating at various sites. More qualitative observations suggest that the same signal also controls skin blood [262].

A final aspect of Fig. 12.1 refers to the relative weights of both regional temperatures in generation of the compound core temperature signal. The symmetry of the contour lines on either side of the diagonal shows that **trunk** and **head temperature signals** make **equal contributions** to the control of HP and REHL. This statement is valid when both regions have identical temperatures and the body core is a thermally homogeneous entity – a qualification which is often, but not always, met, for example in the case of selective brain cooling.

12.1.1
Selective Brain Cooling: a Special Pattern of Core Temperatures

When artiodactyls or felids develop hyperthermia, the temperature of the brain often rises less than the temperature of the rest of the body core. The mechanism underlying selective brain cooling (SBC) was described in detail in Chapter 8. An important feature is that SBC is not mandatory during hyperthermia but can be implemented, fine–tuned or even switched off. The variable operation of SBC has consequences on the internal temperature field and the compound core temperature signal.

The contour lines of Fig. 12.2 (taken from Fig. 12.1) predict that REHL is 1.2 $W \cdot kg^{-1}$ when Thead and Ttrunk are 40.4 °C. However, Tbrain and Ttrunk do not develop along the line of identity during hyperthermia – at least not in a laboratory environment. In a large series of experiments, average Tbrain was 39.2 °C, when Ttrunk was 40.4 °C [297]. On the basis of Fig. 12.2, the implementation of **SBC** has two effects. The first is that it **inhibits panting** and reduces REHL from 1.2 to 0.8 $W \cdot kg^{-1}$ at 40.4 °C, that is, the animals give up one third of their potential heat loss [295]. The second is that, in hyperthermic animals employing SBC, **trunk temperature contributes more than brain temperature** to the activation of heat loss mechanisms [297].

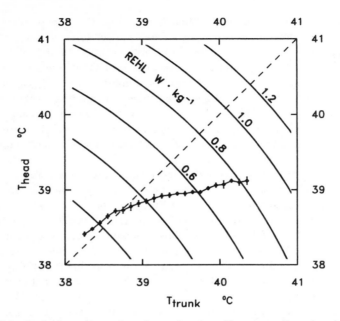

Fig. 12.2. Selective brain cooling and respiratory heat loss. The contour lines show levels of respiratory evaporative heat loss by panting (*REHL*) in goats, when the temperatures of the head (*Thead*) and the trunk (*Ttrunk*) were manipulated independently of each other. The *symmetry* of the contour lines implies that afferent temperature inputs from the head and the trunk contributed equally to heat loss, when the temperatures of both regions were equal. The *single continuous line* shows means and SEMs of Thead plotted vs. Ttrunk when only Ttrunk was manipulated and the head was allowed to adopt its own temperature. Owing to SBC, Thead rose less than Ttrunk. The *line* is based on 55 h of data collection in three goats. (After[260], with permission)

SBC occurs regularly in the **laboratory** when Tbrain of resting animals exceeds a threshold temperature near 39 °C. Two studies addressed the question of whether this is also the case in free–ranging animals in their **natural environment**. The species were black wildebeest and springbok, medium–sized and small antelopes indigenous to the open grassland of southern Africa [267,339].

Figure 12.3 shows the relationship between Ttrunk and Tbrain in four wildebeest while being undisturbed. The line connecting the means of Tbrain intersects with the line of identity near 39 °C; this was the mean threshold of SBC. It agrees with laboratory results in tame artiodactyls (Fig. 12.2). Not seen before, however, was the large variability in the 38–39.5 °C range of Ttrunk, which encompassed the temperatures normally adopted by the animals: the brain was either warmer or cooler than the trunk. Thus, there was **little evidence for a tight thermal control** of SBC. The animals just showed a tendency to employ SBC if they experienced – rarely in their everyday activities – mild hyperthermia. However, SBC was implemented, on occasion, also in the normothermic range of Ttrunk, pointing to a **non–thermal** component in the control of SBC.

BW #2, BW #3, BW #4, BW #5

2–min samples, 4 × 14400

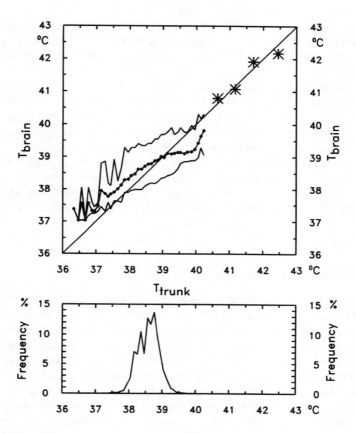

Fig. 12.3. Brain temperature (*Tbrain*) as a function of trunk temperature (*Ttrunk*), and frequency distribution of Ttrunk. 20 days of continuous measurements in four free–ranging wildebeest. *Top* Means, minima, and maxima of Tbrain for each 0.1 °C class of Ttrunk; *bottom* Relative frequencies at which the 0.1 °C classes of Ttrunk occurred; *Asterisks* Highest Tbrain and the corresponding Ttrunk that occurred when the animals were chased and recaptured at the end of the 20–day period. (After [267], with permission)

The recapture of the animals at the end of the observation period involved a high–speed chase of some minutes' duration, and resembled the situation of an animal pursued by a predator. During and after the chase, both Ttrunk and Tbrain rose precipitously, exceeding 42 °C in one animal. The single symbols in the upper panel of Fig. 12.3 show the highest Tbrain reached and the corresponding Ttrunk of the four animals; they straddle the line of identity. Apparently, during the severe hyperthermia of a chase, the animals abandoned SBC. Similar observations were made in springbok: SBC often occurred at low T_{trunk} when the animals were resting or moderately active, and was switched off during spontaneous bouts of intense activity in spite of high T_{trunk} [339].

If one projects these results on natural situations of free–ranging animals, the scenario could be as follows. During rest and moderate activity, the implementation of SBC and subsequent inhibition of panting engages the thermoregulatory system in a sort of **economy mode**, and the expenditure of water for evaporative cooling is reduced. During strenuous exercise, however, which normally occurs only in life–threatening situations, it is imperative to activate heat–loss mechanisms to their full capacities. Abandoning SBC, allowing T_{brain} to rise, and thereby activating brain temperature sensors would serve the purpose and permit REHL to increase to the maximum possible at a given body temperature. Thus, the implementation of **SBC** in moderate hyperthermia, and its suppression in severe hyperthermia **adjust the activity of heat loss mechanisms** to the intensity of heat stress [260]. To phrase it in the context of this chapter: SBC can be seen as a mechanism altering the composition and magnitude of the core temperature signal. While T_{trunk} and T_{brain} contribute equally to the core signal for cold–defence mechanisms, this is not always the case for heat defence.

12.1.2
Natural Spinal Cord Warming

There is another case in which temperature differences within the body core have regulatory significance. Several species can make use of two mechanisms to increase heat production during cold exposure: non–shivering thermogenesis (NST) in brown adipose tissue (BAT) and shivering. At least in guinea pig [69] and rat [18], shivering is predominantly induced and maintained by decreasing skin and spinal cord temperatures, while NST responds to skin and hypothalamic temperatures. A main site of BAT is the interscapular region, from which veins lead directly into the vertebral sinuses [474]. Thus, the heat generated in **BAT** primarily **warms the spinal cord** and inhibits shivering. During moderate cold stress, NST is called upon as the first line of defence, which does not interfere with intentional movements. During more severe cold stress, however, the general decrease in T_{core} overrides the local warming of the spinal cord, and shivering is activated in addition to NST.

12.2
Skin and Core Temperatures

The following considerations are restricted to so–called steady–state conditions, meaning that temperatures "changed sufficiently slowly so that the rate of change had no significant effect, in the opinion of the experimenter" [67]. Thus, the sometimes important dynamic responses (Chap. 14) to the rate of change in skin temperature (T_{skin}) are neglected.

12.2.1
Autonomic Effector Mechanisms

Most quantitative results are available for the control of heat production (HP) and cutaneous evaporative heat loss (CEHL) by sweating in human subjects. Consequently, the data base is biased towards conditions of low to moderate stress in which more subtle features such as non–linearities are less apparent than in animal experiments exploring the limits. Figure 12.4 shows data after 2 h exposure to different environmental conditions. The subject dressed in shorts was resting or exercising on a bicycle ergometer. For the calculation of HP, exercise–induced fractions were subtracted from metabolic rate [42].

Both HP and CEHL were linearly related to T_{core}. The **effect of T_{skin}** was to **shift** the **T_{core} thresholds**: the lower T_{skin}, the less T_{core} had to fall in order to initiate an increase in HP. Conversely, the higher T_{skin}, the lower was the threshold beyond which CEHL increased with further increasing T_{core}. The inset shows the data in the format of contour lines [$W \cdot m^{-2}$ = f(T_{core}, T_{skin})]. The nearly vertical course of the lines visualizes the **dominance of T_{core}**: if the effects of T_{core} and T_{skin} are compared on a °C to °C basis, the influence of T_{core} was approximately ten times larger than that of T_{skin} [42,358]. This is valid for the control not only of HP and CEHL but also of skin blood flow in humans [59,434], and more or less generally, of heat–loss mechanisms in other species during sustained thermal stress [150,182,228].

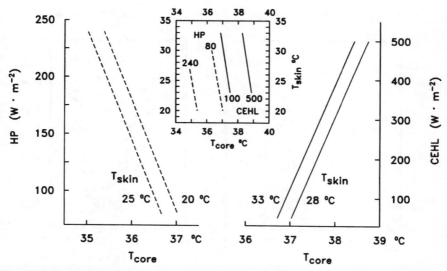

Fig 12.4. Heat production (*HP*), *left*, and cutaneous evaporative heat loss (*CEHL*), *right*, plotted vs. T_{core} at different levels of T_{skin}. The *inset* shows the same data formatted as contour plots. Based on 20 (HP) and 106 (CEHL) experiments in a human subject. CEHL was calculated from sweat rate, assuming complete evaporation. (After data from [42])

The considerations above were restricted to combinations of Tcore and Tskin simulating **natural conditions**, in which **Tskin** is regularly **lower than Tcore**. Within these limits, the control of autonomic effector responses is adequately described by the rule of **linear interaction**: deviations **of Tcore and Tskin** in the same direction contribute **additively** to the central drives, while changes in opposing directions counteract each other. The effect of different levels of Tskin is entirely to shift the threshold of Tcore, beyond which an effector response varies with changing Tcore (Fig. 12.4).

The picture changes to species–specific patterns if **non–sustainable conditions**, that is, Tskin being higher than Tcore, are included. Then the additive interaction in and near the normal range of temperatures appears to present just the central, quasi–linear segment of an essentially non–linear relationship [67]. Figure 12.5 is from a study in which Tcore was manipulated by heat exchangers, while Tskin was clamped by immersing the animal in a water bath of 32 or 44 °C [366].

As in Fig. 12.4, one effect of low Tskin, as compared to the high one, was that the thresholds for increases in HP and REHL were shifted to higher levels of Tcore. However, the **slopes** of HP and REHL over Tcore were steeper at low Tskin. This could still be interpreted in the framework of the additive model: owing to the upward change of thresholds, the zone of proportionality between Tcore and effector activities was shifted to a higher range of Tcore, in which the weight of a unit change in temperature is larger (Fig. 12.1).

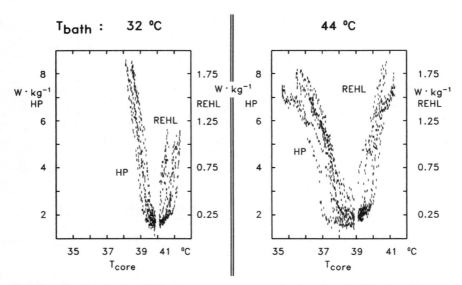

Fig. 12.5. Heat production (*HP*) and respiratory evaporative heat loss (*REHL*) plotted vs. core temperature (*Tcore*): 2 x 4 experiments in a goat at skin temperatures (*Tbath*) of 32 °C (*left*) and 44 °C (*right*). *Symbols* represent 1–min data. The *gaps* between the groups of symbols in each panel indicate ranges of Tcore within which HP and REHL were at resting levels. (After [366])

Figure 12.5 shows that low T_{core} induced strong shivering and a fourfold increase in HP even when T_{skin} was as high as 44 °C. Thus, in agreement with the additive model, cold signals from the skin are not indispensable for shivering — at least in goat. However, the additive model fails when it is applied to a similarly "abnormal", although not uncommon, situation in humans. When a hypothermic subject enters a hot water bath for rewarming, a regular observation is that high T_{skin} suppresses shivering (Fig. 14.4), and HP attains basal levels in spite of low T_{core} [189]. This collides with the results shown in Fig. 12.5 and suggests that, in a species–dependent range of combinations, a **multiplicative** type of **interaction** between T_{skin} and T_{core} yields better descriptions. Its essence is that very high levels of T_{skin} or T_{core} completely suppress cold defence mechanisms, regardless of how low the other temperature is.

A multiplicative pattern has also been found in rabbit, in which hypothalamic temperature (T_{hypo}), manipulated by local thermodes, substituted for T_{core} [480,481,485]. It contrasts with the results obtained in dog, which adhere to the model of additive inputs [208]. A later reexamination of the dog data revealed that the effect of a unit change in T_{hypo} was larger at higher temperature [180], adding a non–linear component, as in Fig. 12.5, to the additive model. The effect is not included in Fig. 12.6.

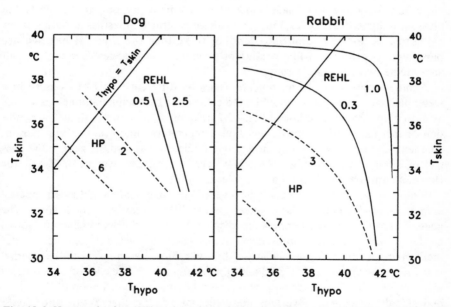

Fig. 12.6. Heat production (*HP*) and respiratory evaporative heat loss (*REHL*) as functions of hypothalamic (*T_{hypo}*) and skin (*T_{skin}*) temperatures in dog (*left*) and rabbit (*right*). The *straight contour lines* for dog indicate the additive interaction of T_{hypo} and T_{skin}, while the *hyperbolic contour lines* for rabbit suggest a multiplicative interaction. *Numbers* next to lines: $W \cdot kg^{-1}$. Sustainable temperature combinations are restricted to those regions of the planes which are below the line *$T_{hypo} = T_{skin}$*. (After data from [208,480,485], with permission)

The results obtained in dog and rabbit (Fig. 12.6) are at variance with each other. The difference is small, and possibly within the range of experimental uncertainty, for those regions of the planes in which T_{skin} is lower than T_{hypo}, and excessive in the range of very high T_{skin}, where the results in rabbit predict the virtual independence of REHL on T_{hypo}. It does not deny the possibly general validity of the multiplicative model, to emphasize once more that the occurrence of its most salient feature is restricted to temperature combinations which are incompatible with life, except for short periods of time.

12.2.2
Behavioural Responses

It is a priori likely that T_{skin} and T_{core} interact also in the control of behavioural responses. The easiest way to address this question is to ask human subjects for the hedonic quality, that is the degree of pleasantness or unpleasantness, of a local T_{skin} at different levels of T_{core}. The inference is, of course, that the subjects, if given the choice, would behave appropriately in order to avoid sensations of unpleasantness. Results of such a study are shown in Fig. 12.7.

The subjects were made hypo– or hyperthermic by cool or hot water baths, then transferred to a neutral bath and asked to rate the sensations evoked by dipping a hand into a set of water baths of different temperatures. The answer of this and other studies was that low T_{skin} is judged unpleasant by hypothermic subjects, but pleasant in hyperthermia [77,356]. The pattern is termed alliesthesia: during internal stress, **peripheral conditions tending to restore the thermal balance are perceived as pleasant**, while conditions potentially further deteriorating the balance are perceived as unpleasant.

Another approach is to manipulate T_{core} by different levels of exercise in a water bath, and then allow the subject to choose the bath temperature he perceives as most comfortable. The line in Fig. 12.7, relating T_{skin} and T_{core} for the condition of thermal comfort, shows that both temperatures interacted, within reasonable temperature limits, in a linear additive mode, as was also the case in the control of HP and CEHL (Fig. 12.4). However, the relative weight of **T_{skin} for thermal comfort** was larger than for sweating [43].

Similar experiments were made in several other mammalian and avian species: animals were given the choice to alter T_{skin} at different levels of local T_{core}. The general answer was that temperature displacements of the hypothalamus, spinal cord or medulla oblongata aroused sensations which motivated the animals to alter their environment, and consequently T_{skin}, in a direction opposite to the internal temperature deviation [438,484,84,309]. In summary, the results of experiments using operant behaviour techniques to determine the relative importance of thermal signals generated in various body regions suggest that **T_{skin} and T_{core} interact in guiding thermal behaviour** basically in the same way as in the control of autonomic responses.

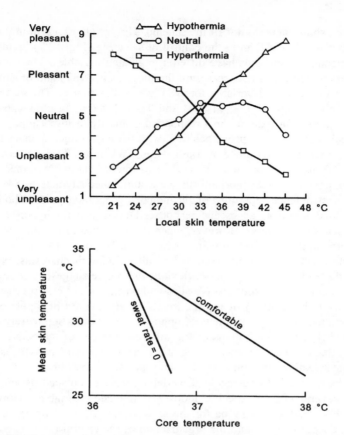

Fig 12.7. Thermal comfort as a function of core and skin temperature in human subjects. *Top* Ratings of local skin temperature between *very pleasant* and *very unpleasant* depended on internal body temperature; 12 experiments in four subjects [356]. *Bottom* A subject exercised in a water bath at different intensities and was asked to adjust water temperature so that he felt comfortable. Regression line based on 21 experiments. In another study on the same subject, the relationship between core and mean skin temperature resulting in *sweat rate = 0* was established. The lines have *different slopes*, indicating that the influence of skin temperature on comfort is larger than on sweating [43]

12.3
The Regulated Variable

In the introduction, the block diagram of the thermoregulatory system (Fig. 1.3) implied that there is **one regulated variable** for all effector mechanisms, reflecting the temperature field of the body by integration of signals from all sensors. This was a charmingly simple and, for a start, useful concept. In its pure form, the concept requires that every thermosensitive site has access to every thermal effector, and that the relative weight of a site is the same for all effector systems.

However, **some facts do not fit** the scheme. The weight of T_{skin} in the control of behaviour was larger than in the control of sweating (Fig. 12.7), implying that the two effector mechanisms have different regulated variables. The same is true for the meshed control of non–shivering thermogenesis (NST) and of shivering in guinea pig: the regulated variable for NST was a function of T_{hypo} and T_{skin}, while shivering followed T_{spinal} cord and T_{skin} [18,69]. In cold–exposed pig, operant behaviour for external heat responded primarily to low T_{hypo}, while the characteristic cold–defence prone position, with the legs retracted under the body, was most frequently determined by T_{spinal} cord. When low T_{hypo} was combined with high T_{spinal} cord, the animals worked for more heat and assumed a spread position: opposing effector mechanisms were simultaneously active [84]. The behavioural heat defence repertoire of rat includes grooming, relaxed body extension, and locomotion as an escape response. During general hyperthermia, the complete set is called upon. However, when discrete sites in the brain stem were locally warmed, 97% elicited only one or two of the responses [415].

A single regulated variable, common to all effector mechanisms, cannot account for these observations. A more flexible alternative suggests a set of specific connections between thermosensitive sites and particular effector mechanisms. In this model, the **integration of inputs**, involving summation of inputs of similar sign and mutual inhibition of inputs of opposite sign, occurs **separately for each effector** [414]. The contributions of the different sources of temperature signals are also effector–specific: comprising all thermosensitive sites for shivering, panting and sweating, and some for NST and specific behaviour.

Thus, the concept of one regulated variable must be replaced by another one, assuming several effector–specific loops and regulated variables. However regrettable this conclusion may be for the sake of simplicity, it opens the way to somewhat deeper insights into the organization of the central nervous component of the temperature–regulating system.

13 The Central Interface Between Afferent Temperature Signals and Efferent Drives

13.1
The Controller: Concepts and Facts

The concept prevailing in the first half of the 20th century was that the rostral brain stem contains two interconnected centres, one for heat loss, and the other for heat conservation [29,282]. It was replaced, at the dawn of cybernetics, by the model of a **single controller**: afferent temperature signals, generated at several sites of the body and converging at successive levels of the brain stem, were thought to reach the posterior hypothalamus, where the integration was completed. In the most rigorous version of the single controller, its function was to process a common regulated variable, to compare it to one set–point and to generate a continuous efferent outflow, proportional to the load error and resulting from the difference between regulated variable and set–point. The graded responses to thermal loads were supposed to be accomplished by effector–specific thresholds: a small load error activates behaviour, a medium one additionally alters skin blood flow, and a large one initiates shivering or sweating [185]. However, the evidence presented in chapter 12 argues against a common regulated variable for all effector mechanisms, and separate regulated variables imply **separate controllers** [427].

Support for this alternative model came from experiments placing small lesions in the central nervous system (CNS) to produce disturbances of **single functions**. In rat, the preferred species for this sort of studies, near–midline lesions in the preoptic area–anterior hypothalamus (POAH) impaired autonomic responses to peripheral thermal stimuli for an extended period of time, while operant behaviour and grooming remained undisturbed [308,414,430,510]. Conversely, lesions in the lateral POAH affected behavioural responses more, and autonomic responses less [431,510]. In cold–exposed rat, skin vasoconstriction required an intact connection, via the medial forebrain bundle, between preoptic area and midbrain, while shivering occurred in spite of the POAH being totally disconnected from the posterior hypothalamus [147]. In other words, the afferent inflow from the skin diverges towards two separate sites in the rostral brain stem: the preoptic area for generating the efferent vasoconstriction drive, and the posterior hypothalamus for shivering. The **spatial separation** of various effector circuits in the rostral brain stem gave the final reason to replace the model of a single controller (Fig. 13.1A) with the currently prevailing concept of **multiple parallel control loops**. Each loop receives its specific set of thermal signals and is serving just one effector mechanism (Fig. 13.1B).

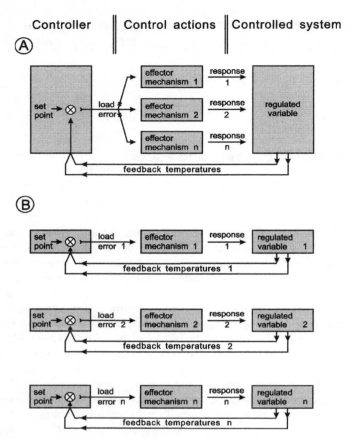

Fig. 13.1A, B. Two models of the central controlling system. **A** A single controller receives temperature signals (feedback temperatures) from all thermosensitive sites of the body (regulated variable). A common load error drives all effector mechanisms that are activated at different thresholds. **B** Each effector mechanism is activated by its specific set of temperature signals, involving separate regulated variables, controllers, and load errors. (After [427], with permission)

Spatial separation of control can even occur within an effector system. Warming the left hypothalamus of rat induced salivation in the left salivary gland, while the right gland remained inactive (Fig. 13.2). Thus, the efferent outflow generated by unilateral warming passes through the medial forebrain bundle to the ipsilateral salivary nuclei of the lower brain stem, without divergence to the contralateral side [277]. For vasodilation, some neural connections between right and left hypothalamus do exist. However, the main efferent pathway in the rostral brain stem is again bilateral, and incomplete fusion of signals from both sides occurs at an infrahypothalamic level, possibly in the medulla oblongata [278,279]. Thus, **within one effector** system, the **efferent drives** to the various body regions can **originate at separate sites** of the rostral brain stem.

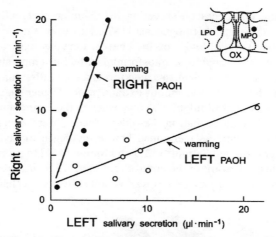

Fig. 13.2. Lateralization of efferent outflow to salivary glands. Unilateral warming of the preoptic area–anterior hypothalamus (*POAH*) of rat generated large saliva flow in the right–side gland (*filled circles*), and low flow in the left–side gland. Warming the left POAH (*open circles*) resulted in the opposite pattern. The weak contralateral responses were caused by spreading of the thermal stimulus to the contralateral side. *Inset* Circles show location of the warming electrodes; *OX* optical chiasm; *MPO* medial preoptic area; *LPO* lateral preoptic area. (After [277], with permission)

Processing of afferent temperature signals into efferent drives can occur also at CNS **levels below the rostral brain stem**. This was suggested as early as 1939 by Thauer [500]. His arguments rested partly on clinical reports: tumour patients, in which postmortem examination revealed complete destruction of the rostral brain stem, had lived till their end without obvious disturbance in temperature regulation. Furthermore, his experiments in rabbit with protracted transections of the lower brain stem showed that the animals maintained near–normal body temperature even during cold exposure. Thauer did not deny the importance of the rostral brain stem in intact animals, but concluded that, provided the lesions develop slowly, lower levels of the CNS could execute functions which are normally performed by higher levels. Later studies confirmed his observations. Kittens decerebrated right after birth showed normal regulatory behaviour and shivering during cold exposure at age 50 days [36]. The best example is the spinalized animal: local cooling of the spinal cord below the level of transection induced shivering in lumbar trunk muscles [87,465] and reduced skin blood flow in the hindpaw [514].

The performance could be expected to improve if higher levels of the brain stem are included. Indeed, lesions at sites of the rostral brain stem in rat did not interfere with behavioural and autonomic responses to local warming and cooling of the medulla oblongata [310]. However, the organization at the brain–stem level is more complex. Shivering and vasoconstriction during cold exposure were abolished by high–level decerebration above the pons, but reinstated by decerebration below the pons [87]. Thus, successive levels of the brain stem exert facilitating as well as inhibiting influences on shivering [6]. Similar observations were made concerning non–shivering thermogenesis [187]. However, they do not invalidate the conclusion that **several levels of the central nervous axis**, including spinal, are at least principally capable of transforming afferent temperature signals into efferent drives to effector mechanisms, and this is a **controller function**.

In an equally far–fetched and ingenious extrapolation, Thauer suggested that low–level components of control might also play roles in the temperature regulation of intact animals – functions that, in the older concept, were assigned exclusively to the rostral brain stem. As mentioned before, evidence in favour of this idea grew in later years, and led Satinoff [427] and Simon [461] to propose concepts of a **hierarchical organization** of the controller. Figure 13.3 shows a simplified version, lumping together the parallel loops of various effector mechanisms.

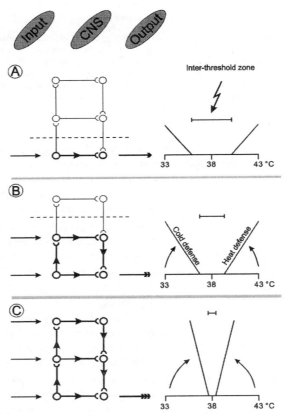

Fig. 13.3A-C. Hierarchy of control levels. **A** The spinal cord is disconnected (*interrupted line*) from the upper CNS, and its limited capacity of transforming thermal inputs into efferent outflow to effector mechanisms results in a wide interthreshold zone and weak responses. **B** Inclusion of the lower brain stem improves performance. **C** The interthreshold zone is narrowest in intact animals, and the slopes of effector mechanisms are steepest. (After [427,461], with permission)

Part A of Fig. 13.3 refers to the isolated spinal cord transforming thermal inputs into efferent outflow to effector organs. However, strong stimuli produce weak responses, resulting in a wide interthreshold zone between cold– and heat– defence mechanisms, and small increases per unit change in temperature. This situation could be called **'broad–band control'**, – a term originally coined to denote a sort of coarse control in intact animals subject to extreme thermal stresses [45].

In Fig. 13.3B, the operating system includes the lower brain stem, which adds more thermal inputs and more connections between afferent and efferent sides, and results in a narrower interthreshold zone and steeper slopes of the effector mechanisms. The intact CNS (Fig. 13.3C) utilizes all inputs and integrating networks, and displays the highest degree of regulatory performance.

The right half of Fig. 13.3 conveys a qualitatively correct picture of the responses of lesioned and intact animals to thermal stimuli, and the diagrams on the left provide a conceivable explanation. However, its remains still to be proved that lower subsidiary levels of control are indeed operative in intact animals. It could also be that all control functions in the intact system are executed exclusively at the most rostral level, and lower levels take over only after more rostral ones were disconnected. Thus, the attractive concepts of spinal, medullary and higher controllers arranged in hierarchical order, with the rostral brain stem as the highest and final level of integration [47,427,461], must remain largely conjectural, as long as they are based entirely on results of lesion and transection experiments.

13.2
Neuronal Models

Models serve the purpose of condensing complex phenomena to comprehensible terms and must simplify: when the complexity of a model approaches that of reality, it can become as difficult to understand, and therefore loses much of its usefulness [46]. It should contain as few elements as possible and, ideally, blend neurophysiological observations with insights gained from the interaction of body temperatures in the control of effector responses. The development of models of central nervous circuits advanced at a time when the technique of recording from single neurons in the rostral brain stem was just evolving. However, the first studies revealed a bewildering variety of individual patterns, whose integration in a comprehensible network of neurons was seemingly impossible [184]. Thus, a class of neuronal models developed that was derived, despite its appellation, mainly from control theory, input–output relations, and synaptic interference studies, taking single neuron data at best as supporting evidence. A neuron in these models does not represent an anatomically defined entity, but rather the behaviour and function of a class of neurons, which may be accomplished by an unknown number of cells arranged in a cascade–like, serial structure [178]. The models' design follows the paradigm of small networks serving motor functions of the spinal cord, which limits the types of elements and keeps the models simple.

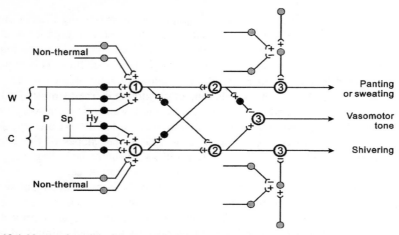

Fig. 13.4. Neuronal model of the control of body temperature. Excitatory warm and cold signals from all thermosensitive sites of the body (*P* periphery; *Sp* spinal cord; *Hy* hypothalamus) are summed in two lines of 1st–order neurons (*1*). 2nd–order neurons (*2*) receive excitatory inputs from their own line, and inhibitory inputs from the complementary pathway. The efferent pathways to effector mechanisms originate at 3rd–order neurons (*3*). (After [47], with permission)

Figure 13.4 shows a model that Bligh has synthesized from various approaches [47]. Warm signals from skin, spinal cord and hypothalamus (and other thermosensitive sites) are arranged in parallel and excite a **first–order** warm–integrating neuron. An analogous set of cold signals from the same sources arrives at a first–order cold–integrating neuron. Both excitatory pathways continue to **second–order** neurons which receive mutual inhibition from the complementary pathway. The function of the second–order neurons is to process opposing temperature signals from different sources (for example, cold skin and warm core), and to ensure that antagonistic effectors like shivering and panting are not simultaneously active. The second–order neurons of the warm–line and the cold–line project on a **third–order** neuron, which is the origin of efferent drives for vasomotor tone and behaviour. Parallel projections pass to other third–order neurons at which the efferent pathways to evaporative heat–loss mechanisms and shivering originate. Third–order neurons receive inhibitory signals to establish an interthreshold zone, in which neither panting/sweating nor shivering are active, while vasomotion (and behaviour) are continuously operating. The network is complemented by non–thermal inhibitory and excitatory influences addressing the first–order neurons.

Originally, the model was designed to provide a template for interpreting the results of **synaptic interference** studies, following observations that injections of transmitter substances like noradrenaline and 5–hydroxytryptamine into the cerebral ventricles of cat and other species produced changes in body temperature. However, later studies revealed that the responses to all substances varied greatly with species, environmental conditions, and site and mode of administration [201]. To date, no consensus has emerged on the possible thermoregulatory functions of the transmitter substances mentioned above, and the same holds for brain peptides [91]. Thus, the model shown in Fig. 13.4 has lost its purpose insofar as it consisted in sketching a general pattern of putative transmitter effects — such a pattern has not been detected. However, the model is probably still a correct pictorial description of the basic functions any conceivable nervous interface between afferent temperature signals and efferent drives must perform in order to fulfil its purpose: summation of thermal inputs of equal sign, mutual inhibition of inputs of opposite sign and integration of non–thermal inputs to establish thresholds. It is essentially a block diagram of indispensable controller functions which are labelled neurons. After non–substantial modifications, the model could accommodate also the concepts of parallel loops and hierarchical organization.

Before leaving Fig. 13.4, it should be noted that, by arranging skin and core temperature sensors as parallel sources, the model treats core temperature sensors implicitly as primary neurons. This is relevant to a difficult problem with modelling: to account for **non–linearities of input–output relations** outside the normal range of temperature combinations. A simple way would be to assume that the relationship between temperature and signal generation of core temperature sensors is non–linear [263,366]. However, the apparent linearity between temperature and responses in single–site thermode experiments occasioned others to look for alternatives [179,462]. One is to assume that the temperature sensitivity of the hypothalamus depends on skin temperature: the lower the skin temperature, the more the hypothalamic network of neurons responds to variations of its own temperature [180,480]. Such models dispense with primary, receptor–like sensors in the spinal cord and the brain. Instead, a network of central neurons is vested with the capability of transducing temperature deviations into inputs to the controller.

In contrast, a recent model retains primary cold and warm sensors in the rostral brain stem and adds the property of unspecific, non–sensory thermosensitivity of neurons involved in transmission and processing of temperature signals [466]. The model was primarily derived from experiments in birds, but can account also for observations in mammals, which are otherwise difficult to explain [397]: cooling the anterior hypothalamus generated shivering (by acting on putative specific sensors), while cooling the posterior hypothalamus reduced it (by inhibiting the transmission of signals in putative integrative neurons). Assuming appropriate coefficients for both modes of thermosensitivity, at least some non–linearities can be accounted for [259].

Neuronal models bear a distinct degree of arbitrariness, which is most obvious in the important question of **primary sensors vs. network** as the source of central thermosensitivity. At present, none of the options can be discarded, and their coexistence is vivid testimony to the problems of analyzing central nervous functions primarily on the basis of input–output relations. Models are often stated to be useful for teaching purposes. Thus, the spectrum of neuronal models may convey what appears as the **presently conceivable alternatives** in the design of central controllers. However, models are also useful for phrasing questions to studies of single neurons.

One question addresses the criteria for assigning specific functions to real neurons. As mentioned above, other evidence indicates that the control of the various effector systems is organized in parallel loops, and Fig. 13.5 offers a proposal for correlating single neurons with specific loops. The neurons are assumed to be inherently warm–sensitive and, in addition, to receive afferent inputs from the skin and the thermosensitive sites of the body core. Thus, the neurons are supposed to act as **sensors and integrators**: the discharge rates reflect their own temperature and the balance between excitatory and inhibitory signals from the periphery. The neurons were sorted according to their spontaneous discharge rates at normal body temperature of 38 °C. The slowest and the fastest neurons had one feature in common: non–linearity. The functional allotment assumes that a neuron serves those **effector mechanisms** which are **active within the neuron's range of highest thermosensitivity**. Neurons being slow at 38 °C displayed an increasing slope of discharge rate over temperature in the high range. Thus the high–range sensitive neurons would be associated with evaporative heat loss mechanisms like sweating and panting. Neurons of the intermediate category, responding linearly to temperature over the full range, would be linked to skin blood flow and behaviour. Fast neurons, not unlikely to receive the strongest input from the skin and having their steepest slope at low temperature, would be allotted to shivering [52].

It is an attractive proposition that the high local thermosensitivity of a neuron in a certain range could be indicative of its involvement in mediating those effector responses which are active in the same temperature range. However, this does not necessarily imply that this neuron acts also as a local temperature sensor, transducing its own temperature into signals to effector mechanisms: it may have functions on the integrative or output side of the network (Chap. 4).

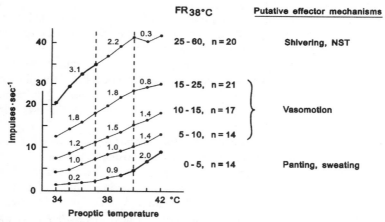

Fig. 13.5. Relationship between neuronal firing rate and range of thermosensitivity. 86 neurons recorded in the POAH of rabbit were grouped according to their spontaneous discharge rate at 38 °C (*FR38 °C*). *Numbers above segments* show the slopes of discharge rate over local temperature; *right* effector mechanisms putatively served by the groups. (After [50], with permission)

The problem of interpretation is by no means specific to the data presented in Fig. 13.5. Single neuron studies of the rostral brain stem in general face the dilemma of recording from an unknown structure with an unknown function [211]. Thus, the inherent ambiguity of current single neuron recordings justifies, in a sense, the persistence of the "neuronal" models mentioned above, whose freedom from real neurons could make them long–living, rather than temporary, substitutes for reality. This is not to deny that their replacement is the most urgent goal of research in this field; it requires the development of criteria for unequivocal assignment of functions to single neurons.

13.3
The Set–Point Problem

The controller in the basic block diagram of the temperature–regulating system, as shown in the introduction (Fig. 1.3), includes a set–point whose value equals that combination of body temperatures at which the activities of all effector mechanisms are at basal levels. The question is how the set–point is generated. In air–conditioning systems, for example, it is realized by a temperature–independent reference signal, with which the feedback signals from the temperature sensors are compared (Fig. 13.6 top, A). The transfer of this simple concept to biological temperature regulation suffers from the scarcity of neurons which produce temperature–independent discharge rates for longer periods.

A more flexible alternative dispenses with a temperature–independent reference and relies on the well established duality of feedback signals (Fig. 13.6 top, B). Of the two types of sensors, one responds with increasing activity to rising temperature, and the other responds with increasing activity to falling temperature (Fig. 13.6 bottom, 2). The load error is generated by comparing the signal rates from both sets of sensors. An increase in body temperature causes the activity of warm sensors to prevail, while the activity of cold sensors dominates at low body temperature. When the signal rates from both sets of sensors balance each other, the load error is zero and body temperature as the regulated variable is at its set–point [341].

The dual–sensor model does not require one set of sensors to have a negative temperature coefficient: it works equally well if the thermal coefficient of the cold–sensor line changes its sign in the afferent pathways so that the signal rates of both sets of sensors increase with rising temperature (Fig. 13.6 bottom, 3). All that is required to establish a set–point are **two sets of neurons with different temperature coefficients** [180]. This condition is fulfilled by all three versions. They can be transformed into each other by rotation of the temperature–signal relationships. Variations of the set–point could be modelled, within the concept of Fig. 13.6, either by parallel shift or change in slope of one set of sensors [210].

Another explanation for the set–point characteristics of the temperature–regulating system was given by Werner [530]. The set–point is thought to be vested in a quadruple of transfer functions: from temperature to signals, from input to the

controller to its output, from controller output to effector activity and from effector activity to the passive system. The transfer functions are arranged in a closed
loop, and any sustained disturbance will finally lead to a combination of operating
points in the transfer functions, at which heat loss and heat production are equal.
The key point of the model is that a biological system could display the features of
set–point control, without requiring comparison of two signals with different
thermal coefficients.

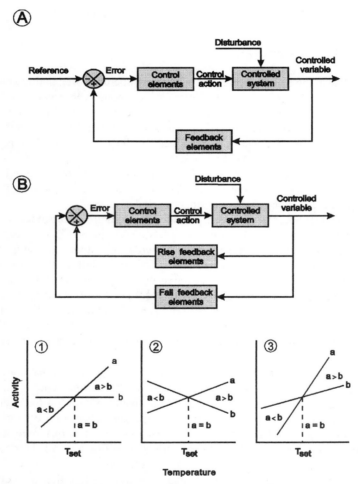

Fig. 13.6. Modelling a set–point. *Top* Control systems with negative feedback, **A** with temperature–independent reference signal, and **B** without reference. *Bottom* Generation of a set–point by
comparing the signal rates from two types of neurons. *1* One type of neuron (*a*) has a positive
temperature coefficient, while the other (*b*) is temperature–independent; *2* one type (*a*) has a
positive, and the other (*b*) has a negative coefficient; *3* both types have positive coefficients of
different magnitude; T_{set} set–point temperature. (After [210,341], with permission)

14 Short–Term Regulation in Various Environments: Inputs and Responses

14.1 The Medium Range

Chapter 12 dealt with the relationships between major deviations in core and skin temperatures (Tcore, Tskin) and the resultant changes in heat production (HP) or evaporative heat loss (EHL). However, animals are not permanently either shivering or sweating: within a range of intermediate temperatures, body temperature is regulated by adjustments of skin blood flow that modify dry heat loss, while the large and costly defence mechanisms against cold or heat are inactive. Figure 14.1 is adapted from Fig. 12.4 and is based on results of a study in a human subject dressed in shorts. The **threshold combinations of Tcore and Tskin**, at which HP or EHL departed from their minima, encompass a zone in which **regulation** was achieved **only by vasomotor control**. Its width was as narrow as 0.25 °C for Tcore, and 3.5 °C for Tskin. Comparable observations were made in goat [258].

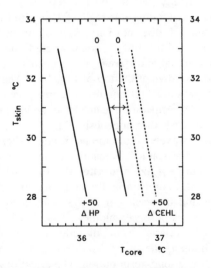

Fig. 14.1. Cold–induced heat production (ΔHP, *solid lines*) by shivering, and heat–induced cutaneous evaporative heat loss by sweating ($\Delta CEHL$, *interrupted lines*) as functions of core and skin temperatures (*Tcore, Tskin*) in a human subject. *Contour lines labelled 0* at the top encompass the central range within which HP and CEHL were simultaneously at minimum levels, and the *thin lines* show the width of the interthreshold ranges. *Lines labelled +50 (W·m⁻²)* refer to increases in HP and CEHL. (After data from [42])

In contrast to the constant conditions in a laboratory, the thermal situation of free–ranging animals is characterized by frequent changes. Resting or moving around, fasting or eating alter HP and subsequently T_{core}, and the 24–h period also of a temperate climate often includes conditions in which T_{skin} or T_{core} is likely to leave the central range. In other words, the everyday life of homeotherms includes short–lasting fluctuations which are small in comparison to what can be experienced in extreme situations, but still large enough to displace body temperature temporarily from the narrow zone of purely vasomotor control.

14.1.1
The Thermoneutral Zone

The interthreshold range as described above is a rather theoretical construct because only vasomotor control was available to the subject as a means of separating the thresholds of cold–induced HP and heat–induced EHL. An approach more pertinent to life outside a laboratory is to replace body temperature with a **range of environmental conditions** and to substitute a less stringent criterion for the departures from the minima: the thermoneutral zone (TNZ) comprises those environmental conditions in which an animal can maintain normal body temperature without activating cold–induced HP or heat–induced EHL. The question is, what options animals have to establish a TNZ larger than suggested by the interthreshold range of body temperatures in a laboratory.

We consider an animal firmly determined to keep body temperature constant at 38.00 °C. It has neither been eating nor really fasting, and its activity is at a low and constant level so that HP does not vary. Skin blood flow is fixed at an intermediate state: the skin is neither fully vasoconstricted nor vasodilated. The heat balance of the fictitious animal in different environments is reminiscent of Fig. 1.1 in the introduction and is shown at the top of Fig. 14.2. The salient feature is that there is just a single environmental condition (corresponding to a **point on the abscissa**), at which body temperature can remain constant without employing shivering and non–shivering thermogenesis (HP+), or panting and sweating (EHL+). Real animals, however, extend the point into a **thermoneutral zone**, and the vasomotor control of dry heat loss is just one of several means.

One way of establishing a TNZ is **to alter the slope of the line relating dry heat loss to environmental conditions**. Reducing the effective body surface area in a cool environment and increasing it in the warmth by postural adjustments (curled vs. spread) serve this purpose; another possibility is to orientate the longitudinal axis of the body in relation to the direction of the sun or wind. If the animal happens to be a human, she or he would certainly change the dress – in furred species, this is only a long–term option. The effect of appropriate behaviour is supported by pilomotion and variation of skin blood flow. Theoretically, if an animal can maintain the dry heat loss (DHL) constant over a range of environmental conditions, body temperature remains also constant within this range and neither HP+ nor EHL+ must be activated.

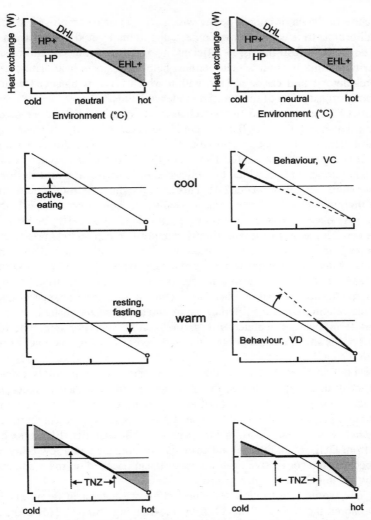

Fig. 14.2. Two ways of establishing a thermoneutral zone. *Top* Heat exchange of a fictitious animal with fixed behaviour and skin blood flow plotted vs. environmental conditions. *DHL* dry heat loss; *HP* temperature–independent heat production; *HP+* cold–induced heat production; *EHL+* heat–induced evaporation. *Middle* Changes in DHL by skin vasoconstriction or vasodilation (*VC*, *VD*) and behaviour (*right*), and in HP by activity and food intake (*left*), in cool and warm conditions. *Bottom* Both adjustments result in a thermoneutral zone (*TNZ*), in which neither HP+ nor EHL+ must be activated in order to maintain body temperature constant

An additional way of transforming the thermoneutral point into a zone is **to alter the level of temperature–independent HP**. In the cooler environment of an early morning or late afternoon, the animal could search for food: the waste heat of locomotion and postprandial excess heat production would both decrease the critical temperature of the environment below which HP+ must be employed to maintain the equilibrium. In a warmer environment, the animal could refrain from eating and sit down for a rest: this is an efficient means of reducing HP and raising the upper critical limit, beyond which E+ must help to dissipate heat. Thus, a perfect thermoregulator could create a TNZ within which DHL is balanced entirely by behavioural adjustments of temperature–independent HP.

Figure 14.2 requires some further remarks. It is based on the premise of constant heat balance (HP = DHL) and could be taken to imply that constancy of mean body temperature is constitutive to the TNZ. However, this is not the case. On the contrary, it is an essential feature of the TNZ that it provides a range of mean body temperature in which the processes of **heat storage and destorage in the body shell** can develop (Fig.7.4). This adds the dimension of time to the concept of thermoneutrality: for periods depending on body mass, mean body temperature may even leave the normal range without HP+ or EHL+ being activated because core temperature does not follow changes in mean body temperature.

Another point is that Fig. 14.2 outlines the responses of an animal to short–term variations of the environment occurring typically within the 24–h period of a climatic season. If **different seasons** are compared, an additional aspect comes into play: by modifying the external insulation (fur), a long–term change in the slope of the line relating DHL to different environments is accomplished, upon which the short–term changes depicted in Fig. 14.2 can be superimposed. The result is that the **TNZ is shifted**, appropriately to the different seasons, to cooler or warmer ranges of the environment.

A third notion refers to the fact that the environments were simply categorized as cold, neutral, and hot. Ideally, the abscissa should be scaled in units of a rational index, comprising the effects of air temperature, air velocity, radiation and water vapour pressure: there is, of course, a passive evaporative component in the heat exchange with the environment (perspiratio insensibilis, EHL), to be subtracted from DHL. Such rational indices exist for humans and a few other species [139]; because the convective heat–loss coefficients depend on mass and shape of the body, the indices are highly specific.

Still, one would like to have an educated guess, at least in terms of air temperature, about the **width of the TNZ** in free–ranging animals. However, even a guess cannot be made: the determination of the onset of HP+ and EHL+ under such conditions is at present technically impossible. This is due to the definition of the TNZ adopted above: if postural behaviour and non–thermal changes in HP are reasonably included as means of extending the TNZ, its limits cannot be determined. All what can be predicted is that it is certainly wider and of more importance than suggested by laboratory experiments.

The usual definition of the TNZ is based on assessing the thresholds of HP+ and EHL+ in restrained and mostly fasting animals; under these conditions, the width of the TNZ is entirely determined by the capacity of variations in skin blood flow and pilomotion to alter DHL. The TNZ defined in this way is primarily of theoretical interest and narrow: for example, 2–3 °C in rat [153]. The TNZ of lean and unclad human subjects in the laboratory is 3 °C or less [183]. A more important practical purpose of determining the range of thermoneutrality in relation to species, age and feeding level is to design housing conditions for farm animals: with reference to the aspect of energy consumption, the range is also termed the **zone of least thermoregulatory effort**, and, consequently, of high productivity [355,517,542]. This definition resembles the one presented above in reference to the TNZ of free–ranging animals.

14.2
Short–Term Exposure to Cold

If the environment is characterized entirely in terms of air temperature, then the criterion for exposure to cold is that the **lower critical temperature (LCT)** is passed. The LCT corresponds to the lower end of the TNZ, implying that an animal has fully utilized its behavioural and vasomotor mechanisms to reduce DHL before HP+ commences. In the laboratory, most but not all species follow this pattern [330], and it appears likely that this is also the case in free–ranging animals, because environmental variations within the nychthemeral cycle usually develop slowly.

Occasionally, however, also a natural environment inflicts rapid changes on unsheltered animals: a cold rainstorm is a good example. Such a situation is more accessible to simulation in a laboratory. Thus, quick transfer of animals from a thermoneutral to a cold room as in the experiments shown in Fig. 14.3 is not totally out of touch with reality, and the animals' responses to the treatment may indeed be similar to those developing in a natural environment.

Figure 14.3 compares **two response patterns** to sudden cold exposure. Data on the left are from sheep with short fleece exposed to +1 °C T_{air}. T_{skin} at the trunk and the legs decreased quickly and HP doubled within 10 min, leading to a 0.3 °C rise in T_{core}. The instantaneous response was followed by a period of decreasing HP and T_{core}, before both variables attained stable levels at min 80.

It is clear that the thermal signals causing the increase in HP originated entirely in the skin: the initial overshoot of HP was a dynamic response to the rapid change in T_{skin}, and its stable elevated level reflected the proportional response to the low level of T_{skin}. Directly comparable observations were made in humans [505]. It is a common feature of species with **minor external insulation** that cold defence mechanisms are driven entirely by **major downward deviation of T_{skin}**: short–haired dog [175], rabbit [150], rat [153], and even shrew [365] showed **increases in T_{core}** during cold exposure. The prime example is nine–banded armadillo [333]. Thus, provided T_{skin} is subject to a major deviation from the thermoneutral level, an overshooting metabolic response is generated and later checked by the rise in T_{core}. It is obvious that free–ranging animals do not often display this type of response: thermal behaviour is directed to avoid rapid and major decreases in T_{skin}.

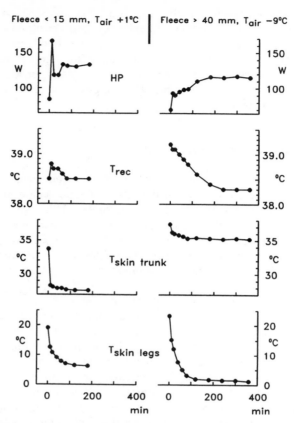

Fig. 14.3. Body temperatures and heat production (*HP*) of sheep during sudden exposure to cold. Fleece depth from 0 to 15 mm (*left*), and > 40 mm (*right*). Both conditions were tested in the same six animals. The *first point of each curve* represents data recorded at +14 °C T_{air} immediately before transfer to the cold room. HP is given per animal. Skin temperatures at the trunk and the legs are means of six and two sites, respectively. (Based on data from [522])

The right side of Fig. 14.3 shows data obtained when the animals were better prepared, by a thick fleece, to cope with major cold loads. Upon exposure to –9 °C T_{air}, T_{skin} at the trunk fell from 37.4 to 35.2 °C after 6 h, – much less than in the short–fleeced condition. The major part of the decrease in T_{skin} occurred in the first 20 min and caused HP to rise; insufficiently, however, to balance the heat loss to the environment. Consequently, T_{core} fell and HP increased further until a stable situation was reached after 4 h.

It is not unlikely that this pattern of response is more pertinent to the situation of larger and **appropriately insulated** animals upon exposure to cold. The point is that the decrease in **T_{skin}** under the thick fleece was too small to generate sufficient shivering. In the case of appropriate insulation, a **decrease in T_{core}** below its thermoneutral level is additionally necessary to induce metabolic responses which are large enough to balance the heat loss to the environment. From the point of view of stability of T_{core}, such a pattern could be considered inferior to that seen in poorly insulated animals. A more important aspect may be energy conservation: appropriate insulation does not just lower the LCT but establishes a response that wastes less energy for temperature regulation [522].

The contribution of decreasing T_{core} must not distract from the **decisive role of low T_{skin}** in maintaining the metabolic response to cold. It is perhaps most obvious in water–bath experiments on humans. The heat loss of subjects resting in water of just a few degrees below thermoneutrality is considerable so that T_{core} drops in spite of shivering [71,475]. An example is shown in Fig. 14.4: during a 50–min period at 28 °C bath temperature, HP nearly tripled and T_{core} (taken at the tympanic membrane) decreased to 35.7 °C. The point, however, is that shivering ceased, in spite of low T_{core}, soon after the subject had left the water bath and was sitting in warm air. Thus, increasing T_{skin} withdrew the skin cold signals for shivering and could even override the core cold signals so that the metabolic response to a definitely hypothermic condition was suppressed [29,374,393].

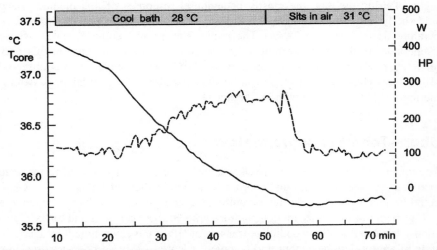

Fig.14.4. Metabolic response (*HP*) of a human subject to a cool water bath and its inhibition by rewarming the skin in air. During 50 min in water of 28 °C, core temperature (*T_{core}*) decreased to 35.7 °C, and HP rose to 270 W. When sitting in warm air, HP returned to normal in spite of T_{core} being still very low. (After [29], with permission)

The question is what levels of T_{skin} are required to exert such inhibiting effects. While no data are available for the experiment of Fig. 14.4, results of another study in human subjects can substitute [189]: after 60 min in a water bath of 10°C, T_{core} was 34.5 °C and the subjects were severely shivering, with HP exceeding 6 $W \cdot kg^{-1}$ (approximately four times resting level). During rewarming, HP remained that high as long as T_{skin} was lower than 33.5 °C. Only when T_{skin} rose further to 37 °C, was shivering completely inhibited in spite of T_{core} being still below 34.5 °C. Thus, rather high levels of T_{skin} (normally associated with overt sweating) are required to suppress shivering generated by such low T_{core}.

Figure 14.3 shows that, in the fully fleeced sheep at –9 °C T_{air}, T_{skin} at the legs fell to lower values than in the clipped sheep at +1 °C T_{air}. However, in spite of the lower temperature at the legs, HP was smaller, suggesting that the decrease in **skin temperature at the trunk** was the primary cause of its initial rise. This is not to deny contributions from skin regions other than the trunk, although no quantitative data are available to document such effects in autonomic cold defence. For cold sensation in humans, the weighting factor of the face is nearly three times larger than can be accounted for by area [99]. The results of another study also ask for caution: cutaneous denervation of the face and trunk did not impair the metabolic response to cold in rat; because T_{core} often increased, it was concluded that cold signals from intact receptors in feet, tail and ears can be sufficient to generate autonomic cold defence responses [190].

Not all species adhere to the pattern outlined above. Warthog occupy the great southern African plateau whose climate is characterized by hot days and cool nights. During the day, the mostly behavioural responses of this medium–sized suid (body mass approximately 60 kg) are fully adequate to prevent body temperature from exceeding 39 °C. The nights are spent sleeping in holes, whose temperature in the early morning is of the order of 20 °C. In spite of this very moderate cold stress, T_{core} was as low as 34 °C, but no shivering was observed [102]. The behaviour of warthog can be categorized as a facet of adaptive heterothermia [311] that is dealt with in Chapter 18.

14.3
Short–Term Exposure to Heat

The upper end of the TNZ is passed when a resting animal is forced to resort to **evaporative heat dissipation** in order to maintain thermal balance. The cause could be increasing ambient temperature, and the term upper critical temperature (UCT) refers to this case: the temperature gradient between skin and air becomes simply too small to allow complete transfer of HP to the environment by DHL, in spite of appropriate behaviour and complete release of skin vasoconstriction. In a natural environment, heat stress in resting animals is mostly associated with a **radiant heat gain** from the sun and ground, adding to HP. In hot deserts, even a convective heat gain is possible if T_{air} exceeds T_{skin}.

The temperature field of the body during heat exposure is characterized by **increase and convergence of** skin and internal body **temperatures** (Fig. 14.5). The core–shell pattern (Chap. 7) is largely replaced by a more or less homogeneous internal temperature field, with the brain as a relatively cooler spot in species employing selective brain cooling (Chap. 8). Thus, temperature signals from all thermosensitive regions act together, and the result is a level of EHL which makes up for the difference between HP and heat gains from the environment on one hand, and eventual convective and conductive losses on the other. T_{core} becomes and remains constant at an elevated level, provided that two conditions are met.

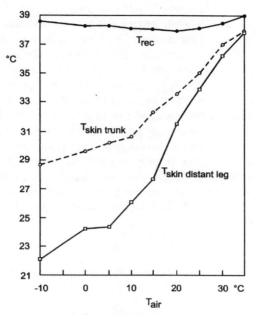

Fig. 14.5. Core (T_{rec}) and skin temperatures plotted vs. air temperature. Mean values of seven dogs, after two hours of rest at each environment. (After [175])

The first is that the environment does not overtax the evaporative heat–loss mechanisms. Generally, their **capacity** is low in small species, and this is the main reason for their preferentially nocturnal life in conditions which include high heat loads during daytime. Larger species in their natural habitat, however, seldom experience purely external heat loads which cannot be coped with by panting or sweating. Their key to thriving in a hot environment is to reduce activity and HP so that the requirements for EHL remain below the limits set by the capacity for sweating or panting. Species with poorly developed autonomic cooling mechanisms compensate for their deficiencies by appropriate behaviour, as pigs do when wallowing in mud [244].

The second requisite for T_{core} to remain in a tolerable range is the availability of water to replenish the loss incurred by evaporative cooling. **Dehydration** generates inhibitory signals to panting and sweating (Chap. 15). The consequence is that T_{core} must rise more in order to compensate for the inhibition by a larger thermal signal to heat–loss mechanisms. As long as T_{skin} is higher than T_{air}, its further rise generates a larger DHL which lessens the demand for EHL. Thus the dehydration–induced rise in T_{skin} and T_{core} is not entirely counterproductive [445,490]. In resting and well–hydrated animals, however, exposure to a few hours of dry heat during the day is rather unlikely to induce a significant degree of dehydration. In a 100–kg animal, a 3% loss of body mass, categorized as low to moderate hypohydration in humans [435], allows an EHL of the order of 3.3 $W \cdot kg^{-1}$ over a period of 6 h, while resting HP is about half that magnitude. Thus the other half is available to compensate for heat gains from the environment.

Exposure to significant external heat stress always results in increases in T_{core} that, according to the Arrhenius–van't Hoff relationship, must increase basal HP and therefore contribute to the heat load. In human subjects, the testable range of T_{core} is small, and so is the effect of T_{core} on basal HP [186]. The range is particularly wide in dehydrated camel, and resting HP at 41 °C T_{core} was indeed 65% higher than at 35 °C, equalling a Q_{10} of 2.1 [451]. The main evaporative heat loss mechanism of camel is sweating, making it unlikely that the metabolic cost of panting contributed significantly to higher HP at 41°C T_{core}. It does, however, in panting animals exposed to **hot and humid** conditions: second–phase panting, the relatively slower and deeper type of breathing, is always associated with increases in HP which deteriorate the efficiency of panting as a means of heat dissipation [162,168].

15 Exercise in the Heat: the Ultimate Challenge

Upon transition from rest to exercise, heat production (HP) increases instantaneously, while the heat–loss mechanisms are still operating at resting levels. Thus HP exceeds heat loss (HL) during the initial stages of exercise. The difference between HP and HL is stored and causes core temperature (T_{core}) to rise. The **increase in core temperature activates heat–loss mechanisms** such as panting, sweating and higher skin blood flow. With increasing HL, the rate of heat storage is attenuated and T_{core} rises more slowly. Finally, as exercise continues, HL matches HP, and T_{core} stabilizes at an elevated level. Thus, on the one hand, the increase in T_{core} is the result of HL lagging behind HP. On the other hand, it links HL to HP: the increase in T_{core} is an indispensable prerequisite for achieving and maintaining a rate of HL which is **proportional to the rate of heat production**. This is not to neglect the input provided by mean skin temperature, and the modulation of sweating and skin blood flow by local skin temperature (Chaps. 10 and 11). During exercise, however, the role of rising T_{core} in activating the heat loss mechanisms is dominant.

An earlier explanation of the increase in T_{core} assumed that exercise raises the set–point of the thermoregulatory system [379]. It was based on experiments suggesting that T_{core} at a given level of exercise is independent of air temperature (T_{air}). However, later studies provided evidence that this not exactly true at low to medium levels, and not at all at high exercise intensity [42,105]. At present, there is nearly general agreement that the steady–state equations correlating sweat rate and skin blood flow with core and skin temperatures during rest hold also during exercise, dispensing with the assumption of a change in set–point [434,486].

In the first paragraph, the scenario was that, as exercise continues, HL increases to match HP, and T_{core} stabilizes at an elevated level. The question is whether this is always the case. Part of the answer is that it depends on the rate of HP. In general, the **maximum metabolic rate** (MMR) during exercise is nearly a constant multiple of ten times resting metabolic rate (MR). This was the result of a study covering many wild and domestic species, and several orders of magnitude of body mass [494]. However, some species or individuals of a species do better than the average: MMR of dog and horse is of the order of 30 times resting MR, and highly trained humans reach 20 times, twice that of sedentary people [526].

The efficiency of transforming chemical energy into mechanical work is about 20% under optimal conditions on an ergometer, and much lower in most natural situations such as trotting or running. The consequence is that at least 80% of MR during exercise is simply HP. In view of the metabolic scope described above, it is obvious that dissipating the waste heat of exercise can pose problems for species prone to engage in high levels of activity. The **maximum rate of heat loss** is determined by three factors: (1) the capacity of HL mechanisms such as sweating or panting, (2) the cooling power of the environment and (3) the internal heat transfer from the exercising muscles to the body surface. If the capacity of all three factors exceeds HP, T_{core} stabilizes at an elevated level. However, if the rate of HP is beyond the limits set by one of the three factors, T_{core} continues to increase.

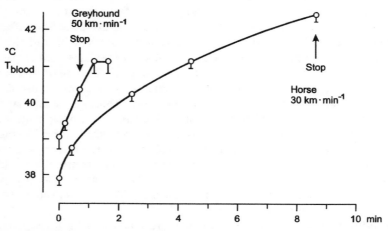

Fig. 15.1. Continuing increase in core temperature at top levels of exercise. In four greyhounds, blood temperature in the pulmonary artery (*Tblood*) rose at a rate of 1.8 °C·min⁻¹ during a 700–m sprint. Estimated metabolic rate (MR) during the run was 35 times basal MR. Six horses ran at 30 times resting MR (90% of maximum) for 8.7 min at 22 °C air temperature. Final T_{blood} in a carotid artery was 42.4 °C. (Greyhound data from [381], horse data from [226], with permission)

The simplest case is elite athlete species exercising at levels near MMR. Except in a cold environment, HP exceeds all limits to HL, and heat continues to be stored. The character of the rise in T_{core} changes: from linking the activity of HL mechanisms with the rate of HP, to reflecting the fact that even the maximally attainable HL is insufficient to match HP. T_{core} may finally reach a level that, alone or together with non–thermal determinants, enforces the cessation of exercise (Fig. 15.1). Exercise near MMR is not restricted to human athletes participating in sprint competitions, and domestic animals incited to chase a mechanical hare or to win a race. Often, hunting is likely to elicit MMR in both the predator and the prey. Cheetah and Thomson's gazelle make a well–known pair: both are sprinters and have top speeds of the order of 70 km·h⁻¹ [493,495]. However, the gazelle sweats poorly, and the cheetah not at all, and both have a rather modest capacity for panting. Necessarily, the chases are of short duration: if unsuccessful, the hyperthermic cheetah gives up after a few hundred m, and both animals, while resting in shade, have time to dissipate the stored heat. **Excessive heat storage** is most clearly seen in species having high MMR but low–capacity HL mechanisms.

Dog, horse and humans excel not just in sprints, but also in travelling long distances at relatively high speed. Under these conditions, storing heat is no feasible option, and the rate of exercise must be adjusted to the cooling power of the environment and the capacity of HL mechanisms. Dogs rely on **panting** and can maintain exceptionally high rates of ventilation, providing substantial evaporative cooling (Chap. 6). In a study on two small dogs running for 15 min at 15 km·h^{-1}, MR was ten times the resting level, and about one half was dissipated through the respiratory tract and the tongue, amounting to 15 W·kg^{-1} [497]. Other orders like artiodactyls have generally less potent panting mechanisms. In goats exercising for 60 min at 42 °C T_{core} and approximately 50% of MMR, respiratory evaporative heat loss was of the order of 3 W·kg^{-1} [123]. While this was a sub–maximum value, it is clear that goat and other predominantly panting artiodactyls cannot compete with dog, and even less with species endowed with a greater capacity for sweating. This is illustrated in Fig. 15.2, contrasting the responses to exercise of the poorly sweating Thomson's gazelle and the profusely sweating eland.

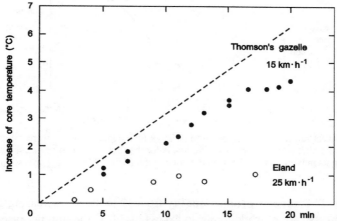

Fig. 15.2. Increase in core temperature in sweating eland vs. panting Thomson's gazelle. The eland (*open circles*) ran at 25 km·h^{-1}, and the increase in core temperature (T_{rec}) after 17 min was 1 °C. The gazelle ran at 15 km·h^{-1}, and T_{core} increased by 4 °C during 15 min. The *interrupted line* shows the predicted increase in T_{core}, if the gazelle had stored its total heat production. It is based on an allometric equation relating metabolic rate to body mass and speed [492]. T_{air} 21–26 °C, relative humidity 17–38%, running area almost completely shaded. (After [493], with permission)

Thus, sustaining high levels of exercise for longer periods of time is, with the exception of dog in moderate or cold environments, the domain of **sweating** species. Trained horses and humans are elite endurance athletes, provided with powerful sweating mechanisms. Humans can produce sweat at rates of 600 to 900 g·m^{-2}·h^{-1}

[42,534], whose evaporation would account for dissipating ten times resting MR. However, air temperature, air velocity and water vapour pressure do not often permit total evaporation, and very fit subjects can exercise for an hour or longer at rates considerably higher than 10 times resting MR. If T_{core} is to stabilize, the difference between HP and evaporative HL must be made up by **convection** and **radiation**, requiring a sufficient temperature gradient between skin and environment. This links the level of sustainable exercise to T_{air}, air velocity and the radiation environment, regardless of the water vapour pressure of the air. Another aspect is that in many natural situations, neither the intensity of exercise nor the environment is likely to remain constant over time. An uneven terrain may alter HP if a runner does not adjust pace, and the sun may dispel a cloud cover, adding a radiant heat load to HP. Thus, it is not without risk to exercise at rates fully exploiting the capacity of HL mechanisms. A rational approach to determine the **safely sustainable rate of exercise** is given by the relationship between T_{core} and sweat rate (Fig. 15.3).

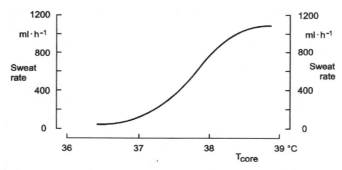

Fig. 15.3. Sweat rate vs. rectal temperature (T_{core}). Five human subjects were exercising at 27–34 °C T_{air} and high relative humidity. The *line* is for the first hour of work and shows sweat rate approaching its maximum near 39 °C core temperature. (After [534], with permission)

Within the 2 °C **zone of proportionality** from 37 to 39 °C, increasing T_{core} generated more sweating, while T_{core} in excess of 39 °C did not, because the maximum sweat rate was reached [104,536]. The same 2 °C zone was found for skin blood flow [377,378] in humans, and data obtained in goat suggest that it is also valid for respiratory heat loss in panting species [366]. Thus the criterion of a safely sustainable rate of exercise is that T_{core} remains within the steep segment of the curve relating T_{core} to HL effector activity, where small increases in T_{core} induce large responses. It provides the physiological basis for heat stress indices which aggregate climatic factors and work rates in order to determine tolerance limits in industrial environments [34].

Beyond the proportional zone, the **internal heat transfer** to the skin becomes increasingly important as the limiting factor. During exercise in a hot environment, muscle and skin compete for cardiac output which cannot increase adequately to meet all demands simultaneously. Nutritional muscle blood flow is determined by the rate of exercise and remains constant, despite the demand of high Tcore for increasing skin blood flow [172,376]. The first line of defence is to reduce blood flow to regions not directly involved in exercise: the splanchnic and renal vascular beds vasoconstrict. However, also the **skin** is subject to **relative vasoconstriction** in a situation of inadequate cardiac output.

The rate of convective heat transfer from the body core to the skin is proportional to the product of skin blood flow and the **temperature gradient between core and skin**. Thus, in spite of sub–maximum skin blood flow, a sufficient heat transfer can still be achieved by a larger gradient if Tcore increases to a higher level [357]. However, in borderline combinations of high exercise intensity and inadequate cooling power of the environment, Tcore could have to leave the zone of proportional control of HL mechanisms in order to establish the necessary gradient (Fig. 15.4).

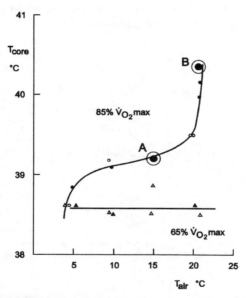

Fig. 15.4. Final core temperature (*Tcore*) after 1 h of exercise at different air temperatures (*Tair*). Two subjects (*open* and *closed symbols*) were exercising at 65 (*triangles*) and 85% (*circles*) of their maximum metabolic rates (*VdotO2 max*). Tcore was measured in the rectum. See text for further discussion. (After [105], with permission)

In Fig. 15.4, final Tcore after an hour of exercise is plotted vs. Tair. Two very fit subjects (MMR was 15 and 19 times predicted BMR, respectively) worked at 65 or 85% of MMR. At the lower work rate, both had no thermal problems within the test range of Tair. At 85% of MMR, however, Tcore left the zone of proportional control when Tair exceeded 15 °C. The study does not provide skin blood flow data. However, in view of the evidence presented below, it appears likely that at 85% of MMR, a relative cutaneous vasoconstriction reduced skin blood flow to an extent which required a large temperature gradient between core and skin for sufficient internal heat transfer. At 15 °C T_{air} (point A), T_{skin} was relatively low, and T_{core} could be maintained at the border of the zone of proportional control. At 21 °C T_{air}, however, T_{skin} was higher, and the internal heat transfer required T_{core} to increase to a very high level (point B).

In such a situation, T_{core} may still be stable for a limited period of time (Fig. 15.5). However, this is not the result of proportional control; it rather reflects a **passive equilibrium between heat production and heat loss** which is very labile for two reasons. First, a small further rise in T_{air} can disturb the balance and directly drive T_{core} to the level of heat stroke, without buffering by physiological responses. Second, skin blood flow during exercise retains its high sensitivity to T_{skin} (Fig. 15.9). Thus a second, potentially deleterious effect of increasing T_{air} is that the critical balance between constrictory and dilatory influences on skin blood flow is affected. If venous return and arterial blood pressure cannot be maintained, the cessation of exercise is enforced.

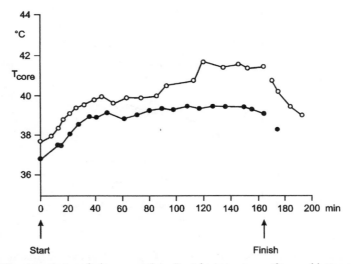

Fig. 15.5. Core temperature during a marathon. Rectal temperatures of two subjects participating in a competition were measured at 9–min intervals. From min 120 to 160, T_{core} of one runner was close to 41.9 °C. Average air temperature and relative humidity were 18.6 °C and 54%, respectively. (After [323], with permission)

An observation common to humans and other species is that **hyperthermia and fatigue** during exercise are closely linked. The level of T_{core} at which fatigue occurs differs between species: around 40 °C T_{core} in rat [136] and highly trained human subjects [152], and near 44 °C trunk core temperature in goat employing selective brain cooling [257]. In such studies, fatigue is usually defined as the reluctance to continue exercise; it is clear that this definition introduces a certain degree of arbitrariness. A subject used to a sedentary life quits at an early stage, at which a trained athlete accustomed to rack himself continues easily even in a laboratory experiment. In competitions, athletes go still further (Fig. 15.5). Thus, the more important question is what mechanisms or functions are impaired by high body temperature so that the motivation to continue declines. In humans and horse, a deterioration of cardiovascular performance [378] and detrimental effects on muscle metabolism [227] were ruled out as causative.

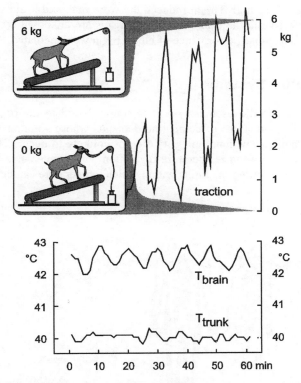

Fig. 15.6. Effects of brain and trunk temperatures (*Tbrain, Ttrunk*) on exercise performance. While Ttrunk was clamped at 40°C, Tbrain oscillated between 42 and 43 °C. When Tbrain approached 43°C, the animal walked more slowly so that *traction* increased. Air temperature 32 °C. See text for further explanation. (After [81])

A study in goat suggests that high body temperature reduces the motivation to exercise by **direct effects on the central nervous system** [81]. In these experiments, another technique was employed to determine the degree of fatigue (Fig 15.6). The animal, trained to walk at a speed of 3 km·h^{-1} and 18% slope, was connected to the front end of the treadmill by a line, and the traction exerted on the line was measured. The traction increased when the animal walked more slowly and let itself be pulled uphill: this resulted in a considerable saving of energy and was considered indicative of imminent fatigue.

The heat–exchanger technique (Fig. 12.1) was used to study the effects of different combinations of brain and trunk temperature (Tbrain, Ttrunk) on performance. No tolerance limit of Ttrunk could be determined: even 43.5 °C Ttrunk did not interfere with walking readily for 1 h. However, there was evidence of a limiting Tbrain. Oscillations of Tbrain between 42 and 43 °C were clearly reflected in the traction record: a rise towards 43 °C was followed by increased traction and imminent fatigue. Thus, high Tbrain reduced the work rate, and it did so at levels which were ineffective in the trunk. The mechanisms are unknown by which the adverse effects on exercise performance of high Tbrain were mediated.

15.1
Cardiovascular System

The capacity of the cardiovascular system to supply blood to the organs of the body is limited, and situations of competing demands whose simultaneous fulfilment would violate its functional integrity are not unlikely to occur. This section deals with conflicts between temperature regulation and maintenance of circulation that contribute to setting the limits within which temperature can be regulated.

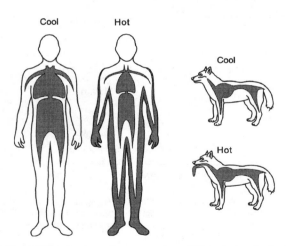

Fig. 15.7. Distribution of blood volume (*grey areas*) on different vascular beds in cool and hot conditions. In a cool environment, skin blood flow is low, and the central blood volume is large in spite of an upright position. In the heat, skin blood flow is high, and a larger part of blood volume is pooled in the distensible venous vessels of the lower body which are subject to high orthostatic pressure. The result is a reduction in central blood volume. Orthostatic problems in hot environment are unlikely to occur in other species: in dog, the tongue as the prominent heat–dissipating organ is normally above the level of the heart, and the effects of mucosal vasodilation on mucosal blood volume are not aggravated by orthostatic pressure. (After [423], with permission)

The competition between thermal demands for skin blood flow and blood flow to other organs is possibly larger in heat–stressed humans than in any other species. First, total skin blood flow during sweating can be of the order of several l·min^{-1} and represent a sizeable fraction of **cardiac output** [422]. The venous section of the skin vasculature is highly compliant, and the constrictor tone of cutaneous veins decreases with increasing skin and core temperatures [544]. Thus, small increases in venous pressure following arteriolar dilation are accompanied by large increases in **skin blood volume**. Second, in contrast to most quadrupeds, the major part of the blood vessels in upright humans is additionally subject to the distending effects of **orthostatic pressure**. The result is pooling of blood in the lower part of the body and a decline in **central blood volume**, with adverse consequences on cardiac filling pressure, stroke volume and arterial pressure.

During supine rest, the skin's requirements for large blood flow could be easily met by an increase in cardiac output. However, there is another option that can become pivotal in situations in which cardiac output approaches its upper limit during exercise: **redistribution of blood flow**. Compared to rest in a thermoneutral environment, splanchnic and renal blood flow are reduced during heat exposure and exercise [422]. Thus, those vascular regions whose functions are not seriously compromised by temporary shortage, "lend" part of their blood flow to the skin. This option is used in humans and possibly all other species, sometimes to the extent that, during rest in the heat, the increase in skin blood flow is accomplished in spite of cardiac output remaining constant [162,173].

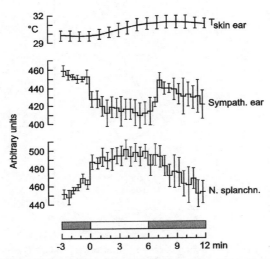

Fig. 15.8. Opposite changes in regional sympathetic activity (*Sympathicus ear* vs. *Nervus splanchnicus*) during skin warming (*white bar*) in anaesthetized and paralyzed rabbits. Decreasing sympathetic activity in the ear was followed by rising skin temperature (*T$_{skin}$ ear*). Skin vasodilation and splanchnic vasoconstriction were caused directly by the thermal input from the skin, because the baroreceptor system was eliminated by denervation and vagotomy. Air temperature 26–29 °C. (After [409])

The reduction in **splanchnic blood flow** is mediated by a regional increase in sympathetic vasoconstrictor activity that contrasts with decreasing activity in the skin (Fig. 15.8). The input signal is purely thermal and does not involve the baro-receptor system of blood pressure control: local warming of peripheral, spinal or hypothalamic temperature sensors induced not only increases in skin blood flow, but also decreases in splanchnic blood flow [468]. In human subjects resting in a thermoneutral environment, the hepatic and splanchnic vessels contain approximately 25% of total blood volume, and most of it is on the compliant venous side [350]. The splanchnic arteriolar constriction reduces splanchnic venous pressure and volume, and helps to counteract the drain on central blood volume caused by high skin blood flow [274,423]. Thus, it is the controlled redistribution of **blood flow and volume** which is part of the circulatory response to thermal loads.

During **rest in an upright position** or **exercise**, the splanchnic vasoconstriction is complemented by a **relative vasoconstriction of the skin** so that, at a given combination of high T_{core} and T_{skin}, skin blood flow is lower than in supine rest (Fig. 15.9). The response is mediated by baroreceptor reflexes and a higher background level of general sympathetic activity so that a larger thermal input, that is, a higher T_{core}, is required for a certain level of skin blood flow [275,317, 434].

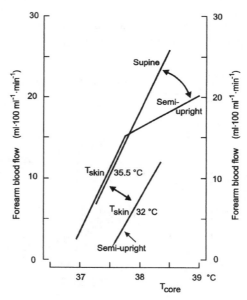

Fig. 15.9. Forearm blood flow during exercise vs. core temperature (T_{core}) at two skin temperatures (T_{skin}) and positions. Data from three human subjects (*Supine* one subject). At 35.5 °C T_{skin}, the increase in blood flow with rising T_{core} was attenuated above 15 ml·100 ml^{-1} ·min^{-1}, when the venous return was compromised by a *semi–upright* position. Note the effect of T_{skin} on blood flow at a given level of T_{core}. Blood flow was measured by plethysmography and is given in ml blood·min^{-1} and·100 ml^{-1} forearm tissue. Its increase was essentially confined to the skin compartment of the forearm. (After [359], with permission)

The compromise is tolerable provided T_{core} remains in the zone of proportional control: an adequate heat transfer from the core to the skin is accomplished by a relatively lower skin blood flow times a larger gradient between T_{core} and T_{skin} [357]. Thus, a larger fraction of cardiac output is available to other organs such as exercising muscle, and the detrimental side effects of skin blood flow on skin blood volume are partly and temporarily mitigated.

During prolonged and intense exercise in adverse climatic conditions, however, the shift of blood volume to the skin can finally exceed all regional compensations [423]. Central venous pressure and cardiac filling decrease, and the only way to maintain cardiac output constant in spite of decreasing stroke volume is to increase heart rate (**cardiovascular drift**, Fig. 15.10). It is mediated by baroreceptor reflexes, possibly complemented by a direct temperature effect on the heart [152].

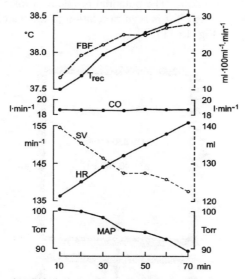

Fig. 15.10. Cardiovascular drift during exercise at 60% of maximum metabolic rate. Core temperature (T_{rec}) and forearm blood flow (FBF) increased during 60 min of exercise. Cardiac output (CO) remained constant, because the decrease in stroke volume (SV) was compensated by the increase in heart rate (HR). Mean arterial pressure (MAP) declined. Mean values of eight subjects. T_{air} 24 °C, relative humidity 40%, air velocity < 0.2 m·s^{-1}. (After [454], with permission)

It is clear that, with longer duration, higher metabolic rate and hotter environment, the cardiovascular drift can become a limiting factor, because heart rate cannot rise infinitely to compensate for the decrease in stroke volume. Thus, cardiac output finally decreases [152], and the work rate must be reduced so that the demands of skin and muscles for blood flow can be met.

It should be noted, however, that the development of major degrees of cardiovascular drift requires intense exercise in tough ambient conditions. The data of

Fig. 15.10 were obtained from subjects exercising on an ergometer in still air. When the experiments were repeated with a fan simulating natural movement through still air at a speed of 16 km·h^{-1} (roughly equivalent to running at 60% of MMR), the changes in stroke volume and heart rate were not significant: with Tskin and Tcore being slightly lower, skin blood flow was just half as large as before, and pooling of blood volume in the skin did not pose problems [454].

15.2
Body Fluid Balance

Evaporative cooling is costly in terms of water consumption. Sweating during a marathon on a warm day can incur a water loss of the order of 5% of body mass that would represent a significant, but not unusual degree of dehydration if it is not averted by intermittent drinking. The potential problem of losing 50 ml water·kg^{-1} body mass becomes apparent if it is seen in relation to the water content of a normally hydrated subject (Fig. 15.11).

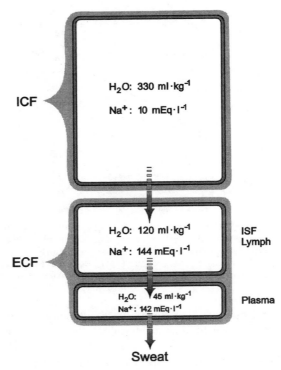

Fig. 15.11. Distribution of total body water and sodium between intra– (*ICF*) and extra–cellular (*ECF*) fluid compartments. Subcompartments of the ECF are interstitial fluid (*ISF*) plus lymph, and plasma. Water content is given in ml·kg^{-1} body mass, and sodium concentration in mEq·l^{-1} fluid. Slowly equilibrating and inaccessible spaces such as bone and cartilage are not included

Neglecting the local environment of the sweat glands, plasma can be considered the immediate source of water for the formation of sweat. However, the interstitial fluid (ISF) and plasma subcompartments of the extracellular fluid (ECF) space equilibrate rapidly by Starling forces: mainly the increase in colloid osmotic pressure, accompanying the withdrawal of water from the plasma, transfers water from the ISF space into the vascular space. Thus, in a dynamic balance, the water used for sweat is taken from the larger store of the entire ECF space.

With prolonged sweating, the decisive question is whether the ECF space can recover part of its loss from the intracellular fluid (ICF) space which contains nearly two thirds of the accessible water. The answer is that it depends on the electrolyte concentration of the sweat [316]. The analysis of two fictitious extremes can clarify the point. The first is that the sweat may consist of **pure water**. Then the osmotic pressure of the ECF must rise, and this provides the driving force for translocation of water from the ICF space to the ECF space. Once the equilibrium is reached, the water used for sweat is delivered by **all compartments in proportion to their initial volumes**. In this ideal case, the water loss of 50 ml·kg^{-1} body mass would be evenly distributed. Because the total water content is about 500 ml·kg^{-1}, the volume of each compartment would shrink by 10%.

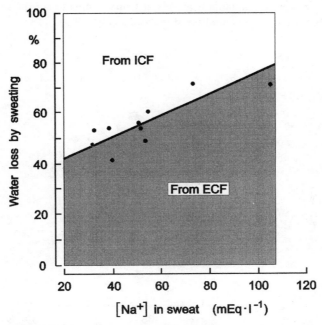

Fig. 15.12. The sodium concentration of sweat determines the contributions of the intra– (*ICF*) and extra–cellular (*ECF*) spaces to the total loss of fluid volume by sweating (100%). Ten human subjects were dehydrated (2.3% loss of body mass) by exercise in dry heat. Data used for calculations were obtained after 1 h of rest in a thermoneutral environment. (After [382], with permission)

The alternative scenario is **isotonic sweat**: its electrolyte concentration equals that of the ECF. In that case, sweating continues to drain **exclusively the ECF** space, because no osmotic gradient exists between the ECF and ICF spaces which could generate a flow of water into the ECF space. It follows that the water loss of 50 ml·kg^{-1} would reduce the ECF space by 30% [50 ml/(120+45 ml)], while the ICF space would contribute nothing to the formation of sweat. The sweat of healthy humans can neither be pure water nor isotonic. However, its electrolyte concentration is variable, and so are the relative contributions of the ECF and ICF spaces to the total loss of volume (Fig. 15.12).

The fraction of total water loss supplied by the ECF space increases linearly with the sodium concentration of the sweat. What matters most is the proportional **reduction in plasma volume** [382]. It adds significantly to the other causes of the decrease in central blood volume, whose detrimental effects on cardiac filling, cardiac output and arterial blood pressure were discussed above, or to phrase it positively, producing a more dilute sweat partly protects the plasma volume and helps to contain the negative side effects of sweating on the cardiovascular system. The production of dilute sweat is a characteristic sign of adaptation to heat. However, adaptation has another and equally important effect: the capacity for sweating is greatly enhanced. While this is clearly beneficial from the point of view of the maximum attainable heat loss, it tends to counteract the effect of adaptation on the sodium concentration of the sweat (Fig. 18.6).

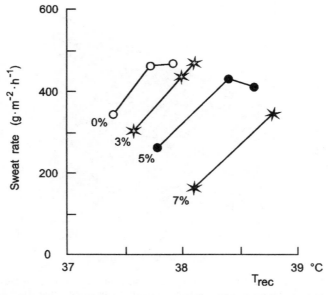

Fig. 15.13. Sweat rate vs. core temperature (T_{rec}) at different levels of dehydration (water loss in % of body mass). Means of eight heat–adapted subjects exercising at moderate intensity in a hot (49 °C) and dry (RH 20%) environment. (After [435], with permission)

In all species, **dehydration inhibits sweating** [35,261,445,490]. This does not necessarily imply a reduced rate of sweating. However, **dehydration requires that core temperature must rise more** in order to attain a level of sweating which is necessary to maintain thermal balance at a given combination of environment and work rate: for example, 400 $g \cdot m^{-2} \cdot h^{-1}$ were produced by euhydrated subjects at 37.4 °C T_{core}, but at 7% dehydration, the same sweat rate required that T_{core} increased to 38.7 °C (Fig. 15.13). In view of the link between sweating and active cutaneous vasodilation in primates (Chap. 10), it is a priori likely that inhibiting effects of dehydration occur also on **skin blood flow**; this was confirmed in resting humans [133,488]. However, also **muscle blood flow** was found to be reduced by dehydration. It was the consequence of decreases in cardiac output and arterial blood pressure occurring near the end of exhaustive exercise [151].

The mechanisms linking dehydration and inhibition of sweating are not yet fully understood. Some evidence suggests the involvement of **baroreceptors** in low– and high–pressure segments of the circulatory system which could signal a reduction in plasma volume to the central regulatory circuits [132]. Other evidence points to osmotic effects: human sweat is hypotonic, and **plasma osmolality** increases in proportion to the loss of free water from the body [382]. The higher osmolality could act directly on the sweat glands, or centrally via osmoreceptor signals relayed to the thermoregulatory circuit [133,373,375,488].

It appears likely that the relative effects of plasma volume and osmolality are again related to the electrolyte concentration of the sweat. The two fictitious extremes mentioned above can be used once more: secretion of isotonic sweat generates a volume error signal but no osmotic, while secretion of pure water would affect both parameters. Horse is a good example: equine sweat is hypertonic [286], and a 6% dehydration induced a 4% increase in osmolality (likely to be attributable to respiratory water loss), but a 23% reduction in plasma volume [145].

By the same token, it could be argued that the **inhibition of panting** during dehydration [490] should be clearly related to an increase in osmolality. As long as no saliva drips from the mouth, panting involves no loss of electrolytes. Evidence in favour of a prevailing role of osmolality in panting species comes from experiments in dehydrated dog: restoring blood volume to the euhydrated level did not remove inhibition of panting, while the uptake of water by drinking returned osmolality and panting to normal [16]. The situation is less clear in rat using mainly **salivation** for evaporative heat loss: the dehydration–induced inhibition correlated better with plasma volume than with osmolality [236]. Upon rehydration, however, osmoregulation had priority over plasma volume regulation [383]. The discrepancy is possibly related to the fact that salivation, in contrast to panting, includes a significant loss of electrolytes.

The role of osmo– and baroreceptors in thermal dehydration extends beyond the inhibition of panting and sweating. An obvious defence against the adverse effects of dehydration on body fluid balance is to curtail the loss of water through the **kidneys**. Urinary output declines in proportion to thermal water loss, and this is caused partly by the osmo– and baroreceptor–mediated increase in antidiuretic hormone which promotes the reuptake of water in the collecting ducts. The second factor is renal vasoconstriction that, as part of the blood flow redistribution during heat exposure, reduces glomerular filtration rate [473]. The renal vasoconstriction is accomplished directly by higher sympathetic activity at arteriolar resistance vessels, and indirectly by release of renin from juxtaglomerular cells which raises the level of Angiotensin II [434].

The renin–angiotensin system is also involved in mediating the sensation of **thirst** that, together with salt appetite, provides the motivation for correcting the deficits incurred by sweating. The intensity of thirst correlates with plasma osmolality and volume. In humans, however, the low palatability of salt solutions tends to delay the restoration of fluid balance (**voluntary dehydration**): drinking pure water dilutes the plasma and removes the osmotically induced inhibition of water excretion in the kidneys. The entire ECF space, including plasma volume, remains contracted unless the electrolyte deficit caused by sweating is compensated [361]. In a sense, voluntary dehydration illustrates the two principal factors involved in body fluid regulation: the electrolyte concentration of the ECF is governed by factors related to water balance, while the size of the ECF space is primarily determined by its electrolyte content [487].

16 Changes of Set–Point

In a technical control system, the numerical value of the set–point is explicitly fixed. In a temperature–regulating animal, T_{set} as the analogue of the set–point must be operationally inferred: it equals the value of the regulated variable if **all effector mechanisms are at minimum levels**. The definition includes vasomotor and behavioural mechanisms whose minima are difficult to define, and, in practice, determining T_{set} by looking for overall minima meets insurmountable obstacles. However, T_{set} affects the **thresholds** of all effector responses, and this opens the way to infer its displacement: if the threshold temperatures of **cold and heat defence mechanisms** have changed in the same direction, it is safe to assume that T_{set} has changed, and the degree of change can be numerically determined. The stress is on simultaneous effects on heat and cold defence mechanisms: changes of T_{set} must be clearly distinguished from changes of single thresholds associated, for example, with thermal adaptation or dehydration which affect either heat production or heat loss mechanisms. The cyclic variation of body temperature is another criterion. In a constant thermal environment, it can be indicative of factors rhythmically altering T_{set}.

16.1
Ovarian Cycle and Circadian Rhythm, But Not Sleep

The ovarian cycle is a well–documented example of a periodical change of T_{set} in healthy subjects. In the second or luteal phase, body temperature is regulated about 0.5 °C higher than in the first phase. Because the thresholds of shivering, skin vasodilation and sweating were all found elevated in the luteal phase, the pattern fulfils all criteria mentioned above [217,218]. It is mediated by central action of progesterone.

The **24–h rhythm of internal body temperature** in a constant thermal environment approximates a cosine wave, with the maximum at daytime in diurnally active species, and at night–time in nocturnal species. The close correlation between activity and temperature does not imply that the former is the cause of the latter [404]. Both rhythms are independent of each other, but follow, as many others, the master clock in the suprachiasmatic nuclei of the rostral brain stem which is entrained by the light–dark cycle of the environment [305,508]. In humans resting in a constant environment, the circadian amplitude is of the order of 1 °C, and, in women, superimposed on the ovarian cycle. It is accomplished by a small phase difference between the rhythms of resting heat production and heat dissipation [420]. Figure 16.1 shows 24–h cycles of body temperature during rest, and threshold temperatures for sweating and skin vasodilation during exercise. All three were closely coupled.

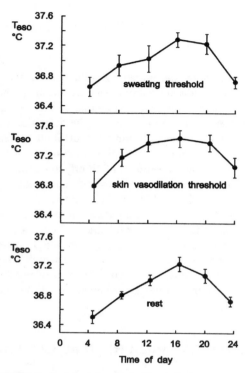

Fig. 16.1. Circadian rhythms of core temperature (T_{eso}) during rest, and core threshold temperatures for sweating and skin vasodilation during exercise. Means ± SEM of 30 experiments in five human subjects plotted vs. time of day. Each experiment was performed on a separate day. Air temperature was 25 °C, and relative humidity 34%. (After [479], with permission)

Of course, the circadian rhythm of motor activity tends to augment the amplitude of the body temperature rhythm. In a constant thermoneutral environment, freely moving tree shrews (body mass 0.2 kg) showed a 5 °C difference between peak and trough [405]. In a thermocline, however, rodents selected lower environmental temperatures during the active phase of the day, and higher ones during the rest phase [63,403]. Thus, thermoregulatory behaviour tends to attenuate rather than accentuate circadian oscillations of body temperature which are linked with alternating periods of activity and rest [428].

The circadian phase of inactivity and low T_{set} is usually associated with sleep. However, **sleep** exerts independent influences on temperature regulation which cannot be fully explained by the change of T_{set}. According to electroencephalographic criteria, its prevalent form is synchronized or slow–wave sleep (SWS). It is interrupted by periods of desynchronization, associated with rapid eye movements (REMS). Figure 16.2 shows, for different stages of arousal, heat production in response to induced changes of hypothalamic temperature in kangaroo rat. SWS lowered slightly the threshold, and more clearly the gain or slope of the response: the increase in heat production per unit change of hypothalamic temperature was reduced. **A feature of SWS is downregulation of body temperature** by lowering the threshold and gain of autonomic cold defence mechanisms [199].

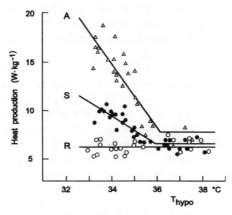

Fig 16.2. Heat production vs. hypothalamic temperature (*T*hypo) during wakefulness (*A*, *triangles*), slow–wave sleep (*S, filled circles*), and rapid–eye–movement sleep (*R, open circles*). Experiments in kangaroo rats at 30 °C air temperature. (After [148], with permission)

The effects were stronger in REMS. The slope was reduced to zero, and even large displacements of hypothalamic temperature remained without response so that the threshold could not be determined. The temporary abandonment of regulation in REMS is not restricted to cold defence. Cats reacted with panting to hypothalamic warming during SWS, but not in REMS [388]. Thus the paradigm of changing T_{set} is not really appropriate for the description of temperature regulation during sleep. Rather, **sleep inhibits all ongoing thermoregulatory activity** [199]. The inhibition is modest in SWS, and extensive in REMS, or to phrase in previously used terms: sleep expands the interthreshold zone, and reduces the gains of autonomic heat and cold defence mechanisms.

16.2
Fever

The concept of T_{set} and its temporary displacement appeals by the convenient way it explains changes of thermoregulatory responses associated with the onset and end of fever. The scheme of Fig. 16.3 describes the sequence of responses triggered by a pyrogenic substance in an animal whose normal body temperature is near 38 °C.

The action of a pyrogen at the beginning of fever is thought to induce a sudden elevation of T_{set}. This places T_{core} below T_{set}, and the resulting **load error** (grey area, horizontal lines) causes sensations of cold, intense skin vasoconstriction and shivering. T_{core} rises and eventually becomes equal to T_{set} (Fig. 16.3B). At this stage, called the plateau phase of fever, intense vasoconstriction, shivering and cold sensations cease: a feverish subject may even feel comfortable, and regulates body temperature by minor changes of skin blood flow as in the normal states A and C. At the sudden end of pyrogenic action, T_{core} temporarily exceeds T_{set} and the initial load error is reversed (grey area, vertical lines). During defervescence the subject feels hot, and skin vasodilation, panting or sweating return T_{core} to T_{set} so that finally state C is reached.

The square–wave change of T_{set} in Fig. 16.3 is certainly a simplification. However, all available evidence supports the assignment of thermoregulatory responses to the different phases of fever [95,207,291]. A crucial point for the concept of a change of T_{set} in fever is that, in the **plateau phase** and absence of thermal stress, neither sweating nor panting is active, and skin blood flow is at an intermediate level. **Elevated T_{core} is regulated as if it were normal**: this distinguishes fever from hyperthermia, in which the same high body temperature at normal T_{set} would strongly activate heat loss mechanisms [319].

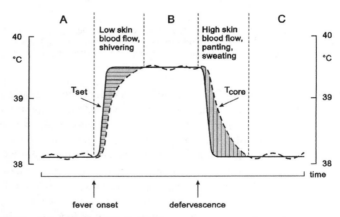

Fig. 16.3A–C. Change of set–point during fever. The *shaded areas* represent load errors created by differences between set–point (*T_{set}, continuous line*) and core temperature (*T_{core}, interrupted line*) at the onset and end of fever. See text for further explanation

16.2.1
Pyrogens and Fever

Fever is part of the defence response to agents which are recognized as foreign by the immune system of the host. Foreign agents (also termed **exogenous pyrogens**)

are lipopolysaccharides (LPS) from the walls of Gram–negative bacteria, muramyl dipeptide (MDP) from Gram–positive bacteria, and viruses all invading the body in the course of infections. However, also partly host–derived compounds such as antigen–antibody complexes (Ag–Ab) are treated as foreign and can induce the **acute–phase reaction**: fever, sickness behaviour, increased release of steroid hormones and production of acute–phase proteins in the liver [337].

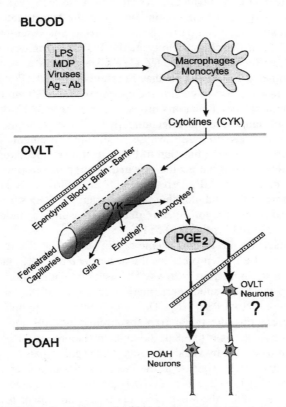

Fig. 16.4. Putative humoral pathway in the development of fever. Foreign agents (*LPS*, *MDP*, *viruses*, *Ag–Ab*) act on immune–competent macrophages and monocytes, and induce the release of *CYK*s. In the *OVLT*, CYKs enter the perivascular space and trigger the release of *PGE2*. PGE2 activates OVLT neurons projecting into the *POAH* neuropile, or reaches the POAH by diffusion. For abbreviations see text

The components of the reaction are **mediated by** a number of **cytokines** (CYK, also termed endogenous pyrogens): large hydrophilic polypeptides or proteins synthesized by the cells of the immune system, mainly macrophages and monocytes, and released into the systemic circulation. The current view is that interleukins 1ß and 6, tumour necrosis factor α and interferon are of prominent importance for the development of the acute–phase reaction. The time courses of production and effects are complex, involving autoamplification, mutual stimulation and inhibition. So far, no single CYK has been identified as the critical agent; the CYK system displays a high degree of plasticity and redundancy [293].

In order to produce fever, the CYKs must communicate with the neuronal circuits of the preoptic area–anterior hypothalamus (POAH) which determine the set–point of the temperature–regulating system. Owing to their size, CYKs cannot pass ordinary capillary walls in the brain. However, the fenestrated capillaries of the circumventricular organs, in particular of the **organum vasculosum laminae terminalis** (OVLT) in the close vicinity of the POAH, permit passage into the perivascular space (Fig. 16.4). There, the CYKs are thought to activate resident monocytes, endothel or glia cells so that **prostaglandin E2** (PGE2) is formed from its precursors and released outside the ependymal blood–brain barrier. PGE2 is thought to stimulate OVLT neurons projecting into the POAH, or to reach the POAH by diffusion and to exert its effect directly on POAH neurons [92,482].

It is the step from CYKs to PGE2, whose details are still enigmatic [39]. However, there is little doubt on the function of PGE2 as the key mediator of fever and the agent ultimately responsible for the **upward change of set–point** [95]. Injections of minute amounts of PGE2 into the OVLT are followed by large increases in body temperature [483], and blocking the formation of PGE2 suppresses the fever response to LPS. In fact, **antipyretic drugs** like aspirin or indomethacin act via the inhibition of PGE2 synthesis [9,511].

A number of facts do not fit the scheme of Fig. 16.4. The rise in T_{core} subsequent to intravenous LPS often precedes the occurrence of CYKs in plasma [38]. The search for a faster **neuronal pathway** between circulating foreign agents and the POAH led to the liver: subdiaphragmatic transection of the **vagal nerve** attenuated the response to LPS [518]. The liver contains macrophagic **Kupffer cells**, possibly presenting the major trap for circulating LPS and other foreign agents. A current hypothesis is that Kupffer cells in response to LPS release a substance which stimulates nearby vagal afferents. The vagal signals are thought to be relayed to pathways which originate at the medullary vagal nuclei and project to the POAH and possibly OVLT, where PGE2 is assumed to be released [38,457].

The substance activating the vagal afferents is still unknown; it could also be PGE2 [40]. Whatever it is, its production requires intact Kupffer cells and involves their stimulation by a complex of LPS bound to fragments of the plasma complement system: elimination of the cells or depletion of the complement abolished the fever response to systemic administration of LPS [40,452].

In conclusion, the role of the proposed peripheral neuronal pathway in the development of fever is still at an early stage of understanding. The humoral and neuronal pathways between foreign agents in the blood and the change of set–point need not be mutually exclusive. Natural fevers last longer than the effect of a bolus injection of LPS, and it is conceivable that the early phase involves peripheral neuronal pathways, while a prolonged fever following a continuous invasion of pyrogen is predominantly based on humoral mechanisms [543].

17   Adaptation to Cold

Thermal adaptation comprises the physiological and morphological changes which reduce the strains of stressful environments. Changes occurring within the lifetime of an organism are termed phenotypic, in contrast to genetically fixed conditions of a species [503]. Phenotypic adaptations enlarge the capacity of existing effector mechanisms, typically as a response to **seasonal changes** of the environment.

With regard to adaptation to life in arctic regions, it must be emphasized that the more important effect of cold is indirect insofar as the long and cold winter limits plant growth and **food supply**. If adequate food is available, the direct effects do not usually pose a major threat to the survival of indigenous species. Cold and starvation are, however, additive stresses, both draining the energy reserves. In severe winter, animals are therefore faced with two conflicting demands: the need to increase energy expenditure for the maintenance of homeothermy in conditions of potentially high heat loss, and the need to conserve energy until the new growth of plants begins [523].

Adaptation to a seasonally cold environment has many species–specific and habitat–specific facets. However, body mass is once more a crude but useful criterion for sorting. **Smaller animals** benefit less from insulation by fur, but can retreat to the microclimate of burrows and enter periods of hibernation or short–lasting torpor in order to save energy. However, they must be able to heat up quickly, for periodic arousal or daily foraging. The possession of **brown adipose tissue** (BAT) for non–shivering thermogenesis (NST) is indispensable to the survival of small animals in the cold. A subsection is devoted to newborn animals: the body mass is at least relatively small, exposure to cold is a major threat and, in many larger species, BAT is present in the perinatal period.

Larger species like reindeer cannot burrow and are exposed to often very adverse weather conditions when seeking food. However, the ratio of surface area to body mass is such that **insulation** by fur is a feasible means of reducing the energy costs of temperature regulation. Provided these animals are well nourished, they are not often forced to increase heat production (HP) just for the purpose of maintaining body temperature in the normal range.

Humans warrant a third section: we are tropical animals, and have adapted to life in a cold environment primarily through cultural and behavioural means. Thus, our genetically determined ability to cope with long–term cold exposure by physiological adaptation is likely to be small.

17.1
Smaller Animals

The general pattern of adaptation to active life outside burrows consists mainly of small increases in resting HP and large increases in **maximum heat production** during severe cold exposure which lower the **cold limit**: the lowest ambient temperature at which an animal can maintain a more or less constant body temperature [411]. A well–studied example is Djungarian hamster (average body mass 25–40 g), inhabiting the Siberian steppe and facing extremes of ambient temperature, ranging from 30°C in summer to –40 °C in winter [425]. In cold tests, the maximum HP of animals maintained at natural winter conditions in central Europe was 75% larger than of animals living in summer at a constant temperature of 23°C. The cold limit in summer was –28 °C, and –68 °C in winter [196].

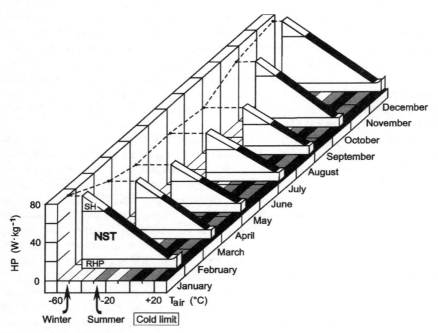

Fig. 17.1. Seasonal changes of capacity for thermoregulatory heat production (HP) depend on the photoperiod. Djungarian hamsters were living throughout the year at the natural Northern Hemisphere light cycle, but at a constant temperature of 23 °C. *RHP* Resting heat production at thermoneutrality; *NST* and *SH* non–shivering and shivering thermogenesis during cold tests; *Cold limit* lowest air temperature at which body temperature could be kept constant in summer and winter. (After [196])

Because the capacity for shivering thermogenesis was the same in both seasons, **BAT** was the main source of the additional HP. In winter, the numbers of BAT cells and mitochondria per BAT cell increase, and the concentration of uncoupling protein per mitochondrium is higher [221]. One environmental cue triggering the morphological and functional adaptation of BAT is of course low ambient temperature. However, at least in Djungarian hamster, the **photoperiod** was responsible for about one half of the seasonal change. Thus, due to the shortening periods of daylight in fall, the thermogenic capacity increased in advance of the colder season, and was further augmented by prolonged or repeated exposure to cold in winter (Fig. 17.1).

The key element in the photoperiodic control of NST capacity is the pineal gland, receiving information on the light perception of the retina and releasing more melatonin when the days begin to shorten. In simulated spring or summer, the effect on the development of BAT could be mimicked by implants slowly releasing melatonin [197]. The fraction of the adaptive growth of BAT which is attributable to intermittent or continuous **cold exposure**, is linked to higher background activity of the sympathetic system. Thus the system does not only control NST during acute cold exposure, but also the capacity of BAT for doing so. The production of triiodothyronine (T3) from thyroxine for local use in BAT and export to other tissues is also greatly increased [221].

In winter, many small non–hibernators **reduce body mass**, and consequently the overall energy requirements [193]. However, mass–specific resting HP increases with decreasing body mass, and heat loss ought to do the same unless the external insulation is improved. Indeed, the winter fur of Djungarian hamster is somewhat thicker and provides more complete coverage: even the walking surfaces of the paws are lined with fur socks. Thus, the thermal conductance in winter was the same as in summer, in spite of a 40% reduction in body mass [195].

Prolonged exposure to cold involves a greatly increased turnover of energy, and **higher food intake** is a necessity in species too small to store sufficient fat in the summer season. A recent hypothesis links cold exposure and episodic feeding in rat: a dip in plasma glucose following the onset of NST is proposed to provide a glucostatic signal to start feeding, and the rise in body temperature subsequent to the period of enhanced NST to deliver a thermal signal to stop feeding. Because body temperature rises more slowly in the cold, duration and size of the meal could be adjusted to ambient temperature [220].

17.1.1
Hibernation

Inside a burrow, a solution to the energy problem is to temporarily abandon homeothermy. Hibernation is a prolonged reduction in body temperature and HP, periodically interrupted by short–lasting returns to normothermy. It occurs in different mammalian orders, and the range of body mass spans from 0.005 to 80 kg or more, if the bear is included. However, the median of the most common species is 0.085 kg [143].

The minimum level of body temperature depends partly on the temperature of the soil in which the burrow is situated. However, at very low soil temperature, freezing is prevented by a controlled increase in HP. Thus **body temperature in hibernation is regulated** as in normothermy, except that the threshold at which HP increases to prevent body temperature from further falling is shifted to a lower level (Fig. 17.2).

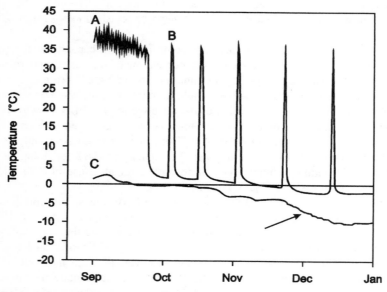

Fig. 17.2. Body core temperature (*A, B*) and adjacent soil temperature (*C*) of a hibernating arctic ground squirrel in its natural burrow. Data of the first half of a 7–month season. *A* Normothermic period preceding entry into hibernation; *B* first arousal to normothermy; *arrow*: when soil temperature decreased from –4 °C to –10 °C, core temperature during torpor remained constant at a minimum of –2 °C, indicating active thermoregulation. (After [55], with permission)

The example shown in Fig. 17.2 is exceptional insofar as other hibernators regulate body temperature above the freezing point of water; it is unclear how arctic ground squirrels prevent freezing at subzero temperatures [21]. Equally unclear is why the torpor is interrupted, at intervals of increasing duration, by short (24 h or less) periods of normothermy [516]. The arousals are expensive from the point of view of energy consumption. However, the cost is tolerable in comparison to the energy saving during the long torpor periods. The median of **metabolic reduction** (HP during deep torpor in percent of resting HP during normothermy) of 36 species was 4% [143]. Estimates of the total energy expenditure during a 7–month hibernation season, including the cost of arousals, yielded savings of the order of 80%, as compared to permanent normothermy [516]. Still, depositing fat in the season before hibernation is the decisive preparation, and the fat fraction of body mass occasionally exceeds 50%. Some species hoard food, to be eaten in the normothermic intervals.

The **entry into hibernation** follows distinct patterns which can be observed even in the same animal on different occasions. In smooth entry, the decrease in body temperature is almost continuous, while in irregular entry, body temperature tends to plateau at intermediate levels for limited periods of time [198]. The second type is often seen in marmot, the largest deep hibernator, and can extend the interval between departure from normothermy and arrival at deep hibernation to a couple of days. This offered the opportunity to study the gradual changes in regulatory characteristics during entry (Fig. 17.3).

The experiments involved local cooling of the hypothalamus by thermodes below its spontaneously attained temperatures at different stages of the entry phase. The threshold temperatures at which cooling induced a rise in HP decreased gradually with deepening hibernation, indicating that **body temperature is regulated also during the entry phase**. Furthermore, the slopes at which HP increased per unit change of temperature grew smaller with decreasing thresholds. A curve joining the symbols of Fig. 17.3 would rise exponentially with a temperature coefficient ($Q10$) of approximately 2.5, and its extrapolation to normothermy suggests that the same **integrative mechanisms operate continuously over the entire range of body temperatures** experienced in normothermy and hibernation [130].

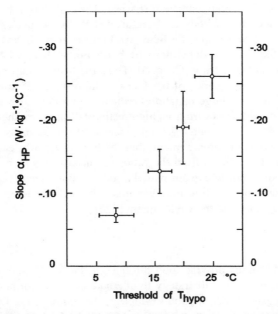

Fig. 17.3. Hypothalamic thermosensitivity in marmots during entry into hibernation. At four stages, the threshold temperature for increasing heat production (*abscissa*) and the response to experimental lowering of hypothalamic temperature (T_{hypo}) below the threshold (*Slope α_{HP}, ordinate*) were determined. During entry, the decreases in threshold and slope were coordinated. Means ± SD of 4–7 animals. (After [130], with permission)

In small hibernators, the decrease in HP during hibernation is larger than can be accounted for by the Arrhenius–van't Hoff effect (Q10) of temperature on biological processes [142,194]. This argues for an additional **specific inhibition of metabolism** by which the centrally determined decrease in the thresholds of metabolic cold defence is executed at the cellular level. One possible link is extreme hypoventilation, preceding and accompanying hibernation. The extracellular pH in deep torpor is maintained at the same level as in normothermy [516]. At low body temperature, however, pH 7.4 represents a severe acidosis, due to the temperature dependence of the neutral point of water [402]. A number of metabolically important enzymes are inhibited by acidosis [321,476].

Prior to **arousal**, respiratory acidosis and metabolic inhibition are reversed by normoventilation. The rapid rise in body temperature is effected by NST in BAT and by shivering. Both are activated by the central nervous system and supported by the temporary positive feedback between HP and body temperature: the higher body temperature, the higher is HP (Q10 effect) which, in turn, accelerates the increase in body temperature.

The **arctic bear** is not really a small animal, and somewhat misplaced in this section. Still, it builds dens in winter and hibernates for approximately 6 months. Two distinct features are that body temperature remains relatively high (32–36 °C), and metabolic reduction is small (30%). However, the mass–specific HP of an 80–kg bear at 35 °C body temperature was not very different from that of small hibernators at much lower body temperature [519]. In 29 species between 0.005 and 4 kg mass, the **mass–specific heat production at 5 °C body temperature was** 0.1–0.2 W·kg^{-1} and **independent of body mass** [142,194,470]. This is in sharp contrast to normothermy (Chap. 5). Thus, the bear's small metabolic reduction is simply the consequence of the fact that its resting HP in normothermy is just 1.1 W·kg^{-1}, and the large metabolic reduction observed in a small hibernator of 0.1 kg body mass follows from its high resting HP in normothermy (6 W·kg^{-1}). If a hibernating 80–kg bear and a 0.1–kg rodent with similar mass–specific HP are exposed to the same cold environment, their body temperatures cannot stabilize at the same level. This follows from the ratios of surface area to body mass which are sufficiently different to explain why the rodent's body temperature approaches that of the environment, while the bear maintains a large temperature gradient between the body core and the environment [225].

17.1.2
Daily Torpor

Some very cold regions provide food throughout the winter and allow daily foraging. Then it is a successful strategy to alternate between normothermy and torpor on a daily basis. Again, a good example is the night–active Djungarian hamster, retreating into burrows during daytime. Figure 17.4 shows one torpor bout.

The general patterns of body temperature and HP in daily torpor resemble those during hibernation (in Fig. 17.4, the ordinate is labelled MR instead of HP, because the normothermic phase included bouts of locomotor activity). However, the high time resolution of Fig. 17.4 unravels a feature which remains undetected when the steady states of normothermy and torpor are compared. The arrow points to the end of the entry phase: the decrease in HP preceded the fall in T_{core}, and HP attained its lowest level while T_{core} was well above its minimum. This is the key argument for the specific inhibition of metabolism in torpor which at least supports the depressing effect of decreasing temperature [194]. Another important result of the same study was that the total **conductance** (Chap. 7) remained constant during entry into torpor. In other words, the animals did not activate heat loss mechanisms during entry (as they would if the fastest way to the torpor level of T_{core} were the goal), but avoided the waste of energy [194]. If that is generally the case, it is somewhat inaccurate to refer, in the context of hibernation and daily torpor, to a change of set–point. The difference between the normothermic state in the cold and torpor is that the internal threshold temperature for the activation of metabolic cold defence is high in the former and low in the latter, while the mechanisms promoting heat loss remain inhibited in both conditions.

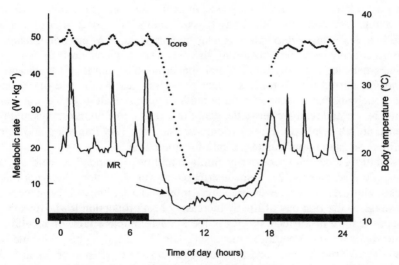

Fig. 17.4. Internal body temperature (T_{core}) and metabolic rate (MR) of a Djungarian hamster during normothermy at nighttime (*black bars*) and torpor at daytime. T_{air} 5 °C. The spikes of MR during normothermy result from bursts of activity. *Arrow* See text. (After [425], with permission)

A major difference between small hibernators and daily torporists in cold climates is that the latter tend to maintain larger temperature gradients between the body and the environment. It results in relatively higher HP during torpor, and consequently smaller metabolic reduction and energy–saving [143,194]. This and the short torpor duration with subsequent rewarming require frequent refilling of the internal energy stores. Thus, regular foraging is both a favourable condition for, and a necessity of, daily torpor [142]. However, neither the requirements for foraging, as dictated by the thermal environment, nor the yield of searching for food in a given period of time will be the same from day to day. It is thus not surprising that daily torpor occurs less regularly than seasonal hibernation [241,425].

A very loose correlation between environmental temperature and the incidence of torpor exists in very small inhabitants of temperate climates. At 0 °C T_{air}, subtropical bats (body mass 10 g) maintained HP at 75 W·kg^{-1} and T_{core} at 35 °C, or resorted to deep torpor, with HP near 12 W·kg^{-1} and T_{core} at +3 °C. A single animal even entered torpor at 25 °C T_{air} [144]. While this laboratory study did not offer any clues to the propensity of individuals for choosing a particular pattern, it is not unlikely that, in temperate climates, torpor is primarily "a response to a limitation of energy, whether it is anticipatory of future deprivation, or a consequence of an immediate withdrawal of food signalling ultimate deprivation" [241].

17.1.3
Everybody Starts Small

– at least relative to the final body size: at birth, the cubs of the polar bear weigh 600–800 g, less than 1% of the body mass in adults. Birth of a litter of 1–3 takes place in a den whose temperature is near freezing. The cubs are blind, naked and have no subcutaneous fat. According to observations quoted in [48], survival of the cubs depends critically on maternal sheltering. The mother curls up and uses her heavily furred legs to press the cubs to her nipples. At the end of a 3–month period, the cubs have acquired a body mass of approximately 10 kg, are well insulated by fur and ready to leave the den. Polar bears, like humans, pigs and many rodents, are **altricial species**: the young are born in a very immature and helpless condition, so as to require maternal care for some time.

Immaturity does not necessarily imply inability to respond to cold stress by autonomic mechanisms like vasoconstriction, shivering or non–shivering thermogenesis. Human neonates, even premature infants of 1 kg body mass, responded to cold stress on the first day of life by increasing heat production [68]. However, the common feature of altricial species is the almost complete **lack of insulation** in the newborn. In this condition, increasing HP in order to cope with cold stress does not make much sense, and thermal behaviour, be it parental or social by huddling, is the more appropriate solution. I cannot phrase it better than Satinoff: "Without heat conservation, increased heat production is worse than useless. The main business of infants is to grow, and it would be a waste to spend energy on achieving a stable body temperature that could not be maintained". [428].

The problems grow in very small species. The body mass of lemming at birth is 4 g. Metabolic responses to cold were detectable on the first postnatal day; they were, of course, futile in isolated animals, and body temperature dropped as in poikilotherms [224]. However, another feature of small altricial species became apparent: the animals survived temporary hypothermia of 14–16 °C body temperature without ill effects. A supreme **tolerance to profound hypothermia** assists very small species in surviving hostile conditions [48].

Other species are **precocial**: capable of a high degree of independent activity from birth, and their distinct feature is adequate insulation. However, the period of relatively temperate environmental conditions in polar regions is short. Thus, fast growth is the first priority for the offspring of ungulates, seals and whales, requiring large intake of fat and protein. In aquatic mammals, the specific problem is to acquire a thick layer of blubber; it may be for this reason that the **fat content of milk** is of the order of 40–50% [41,48].

17.2
Larger Animals

Most of our knowledge comes from domestic animals like sheep and cattle. Both species have a thicker coat in winter, and the **growth of new hair** is entirely controlled by the photoperiod. In sheep, total wool cover and external insulation were the same in two groups kept indoors at 16 °C, and outdoors during a Canadian winter at air temperatures down to –40 °C, respectively [525]. The situation was different in cattle. The total hair cover of animals kept outdoors was twice that of indoors controls, the reason being **reduced hair shedding**. As a result, the calculated lower critical temperature (LCT) at still air decreased by 16 °C [524].

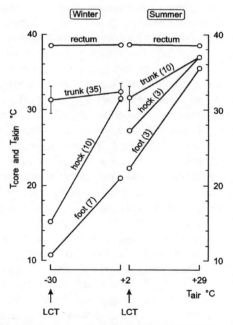

Fig. 17.5. Rectal and skin temperatures of reindeer upon transition from a warm to a cold environment in winter and summer. For animals having a summer coat, +29 °C was warm, and the lower critical temperature (*LCT*) was +2 °C. With a winter coat, +2 °C was warm, and LCT was –30 °C. *Numbers in brackets* fur thickness in mm. (After [273] and [380], with permission)

Similar observations were made in reindeer. In winter, the LCT of animals kept outdoors was more than 30 °C lower than in summer [380], corresponding with increased fur thickness [273]. The insulating properties of the thick winter fur are perhaps best illustrated by comparing the trunk skin temperatures at both LCTs: they were identical in spite of the large difference in air temperature (Fig. 17.5).

The lower LCT in winter needs not be entirely due to improved insulation. It could be assisted by increases in **food intake** and resting HP. Indeed, the appetite of sheep and cattle is larger in winter [523]. However, reindeer [380] and some other cervids [318,345] reduce eating in winter even if food is offered ad lib. In spite of being counterproductive from a purely thermoregulatory point of view (resting HP in winter was just 70% of that in summer), this inherited behaviour presents a perfect adaptation to a natural environment offering plenty of food in one season, and very little in the other. The successful strategy is to store fat in summer, not to suffer from hunger in winter, and to rely primarily on insulation and behaviour to cope with the cold. In Svalbard reindeer, more than 30% of the body mass in autumn is fat [380].

17.3
Humans

The ways of life of aborigines in the central Australian desert and of nomadic Lapps in arctic Finmark did not have much in common before modern technology arrived. However, at night both may have experienced comparable degrees of cold exposure: the aborigines slept naked on the bare ground in spite of nighttime temperature reaching 0 °C, and the Lapps spent the much colder nights of Finnmark in unheated tents. In overnight cold tests, subjects of both ethnic groups shivered slightly and slept well, while controls from temperate climates were often prevented from sleeping by more vigorous shivering. However, the blunted shivering responses in the Lapps and the aborigines were accompanied by relatively larger decreases in core temperature [181,300]. The attenuation of the metabolic response to cold is an example of **habituation**: a diminished response to a repeated stimulus. Most likely, it can develop to some extent also in subjects from temperate climates, provided they experience regular cold exposure at the time when they expect to sleep. The ability to simultaneously sleep and shiver is a sign of cold adaptation in humans [523].

Along the shores of Korea live thousands of breath–holding women divers who each day harvest abalones, seaweeds and the like from the ocean floor. They commence their profession at a young age and continue to dive during many years at water temperatures between 25 °C in summer and 10 °C in winter. Before wet suits came into use in the 1970s, the usually lean divers wore only cotton bathing suits throughout the year and were subject to what was possibly the most intense occupational cold stress in modern times: at the end of a diving shift, lasting 40 min in summer and 25 min in winter, rectal temperature was approximately 35 °C in both seasons.

In laboratory experiments, the diving women tolerated lower water and skin temperatures without shivering than non–diving controls from the same community (Fig. 17.6, top). Thus, one component of their adaptation to repeated cold exposure may again be described as habituation. However, there was another one: at the same thickness of subcutaneous fat, the conductance was significantly lower than in the controls (Fig. 17.6, bottom). Thus **improved internal insulation** was the other component of adaptation in the divers.

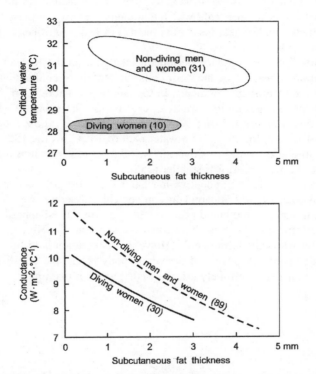

Fig. 17.6. Critical water temperature (lowest tolerable for 3 hours without shivering) and conductance plotted vs. mean thickness of subcutaneous fat in Korean woman divers and non–divers of either sex. Numbers of subjects indicated *in brackets*. (After [230], with permission)

Both components may, in fact, be connected. As long as the inhibition of shivering persists, muscle blood flow can be restrained, and the musculature remains part of the insulating body shell [540]. In addition, the degree of finger vasoconstriction during local cooling was larger in divers than in non–divers, and it was speculated that the countercurrent heat exchange in the limbs had improved [230]. Unfortunately, this idea can never be tested: following the introduction of wet suits, the divers lost their adaptation, and the responses to cold are now indistinguishable from those of the non–diving community [387].

Some occupations require frequent immersion of the hands in cold water while the rest of the body is usually protected from cold by heavy clothing. At least in the 1960s, fishermen [303] and fish filleters [371] used to work for several hours a day with bare hands in water of 10 °C or lower. In experiments simulating the working conditions, these people had higher hand blood flow and finger temperatures than ordinary workers. They did not complain about cold pain, and the blood pressure responses to cold, as an index of general sympathetic activity, were also blunted. The attenuated vasoconstrictor response in the hands can be seen as the result of **habituation to local cold** which maintains dexterity and lessens cold discomfort, while the higher heat loss at the hands has little significance for the heat balance of the well–insulated body.

Thus, apart from the improved peripheral insulation in the Korean women divers, habituation appears to be the most common feature of cold adaptation in humans exposed to natural environments. In the laboratory, habituation should show up as a shift in the **threshold core temperature of shivering**. A recent study showed that this is indeed the case. Young sportsmen were exposed, three times a week during 1 month, for 1 hour to a water bath of 14 °C. At the 13th immersion, core temperature at the onset of shivering was 0.8 °C lower than at the first time, and the degree of discomfort was smaller [255].

Habituation as the leading adaptive mechanism against cold is perhaps unique to humans. Its benefits are apparent in coping with moderate stress of restricted duration: the nomadic Lapp could heat up in the morning by shepherding his reindeer, the aborigine could expose himself to the sun, and the Korean diver could leave the water when she felt too cold. However, habituation is of little use during intense and continuous exposure of the whole body. Thus, the physiological cold adaptation of humans is definitely inferior to that of other mammals.

18   Adaptation to Heat

In hot regions, survival and prosperity of animals depend to a large extent on water: indirectly because precipitation determines the availability of vegetable food, and directly because water is needed for evaporative cooling. Indigenous species have developed behavioural and physiological adaptations to survive on limited food resources and to confine the dependence on evaporative cooling. Again, success of a certain adaptive strategy is linked to body mass, and humans receive separate treatment because their heat adaptation has been studied more intensively than that of any other species.

18.1
Smaller Animals

Several hot and dry deserts are inhabited by a considerable number of rodents which can subsist on relatively dry food and do not require free water. Adaptation includes extreme economy in water expenditure associated with urine formation [439,443] and respiration: the mechanism of recuperation of water on exhalation (Fig. 11.4) is utilized by all species independent of body mass [442]. Rodents lack sweat glands, and it has been emphasized before that small animals, owing to the large relative surfaces, can scarcely afford to use water for evaporative cooling (Fig. 5.1). In view of these constraints, the solution is to limit aboveground activity to the cooler hours of the night. In fact, most desert rodents spend the day in **burrows** and rarely, if ever, expose themselves to the full heat of the day [439].

However, diurnal foraging is compatible with life in the desert. The antelope ground squirrel occupies the North American deserts. Due to its small body mass (100 g), it does not take much time above ground for internal body temperature (T_{core}) to reach the upper tolerance level. The animal then retreats to its burrow for cooling down, and after T_{core} has normalized, returns again to the surface. Only during the hottest hours around noon, does the animal stay in the burrow for a longer period (Fig. 18.1).

A somewhat extreme use of the strategy of living in burrows is made by the naked mole rat, a species of approximately 40 g body mass leading a permanently subterranean life in tropical Africa. Its natural habitat is characterized by two properties. The first is scarcity of food, giving absolute **priority to energy saving**. The second is thermal stability of the environment. Mean burrow temperature is near 31 °C, the daily deviation from the mean is usually less than ±1 °C, and the seasonal variations are unlikely to be significantly larger [28]. The warm environment opens a way to cope with the energy problem. If there is a life–long "guarantee" that environmental temperature never deviates substantially from 31 °C, then it is advantageous to abandon the usual mammalian 36–39 °C range of T_{core} and to **regulate permanently at a set–point just above 31 °C**: the mole rat's resting heat production (HP), at 32 °C T_{core} in its thermoneutral ambient temperature of 31 °C, is considerably lower than that of other mammals regulating at higher levels [70,312].

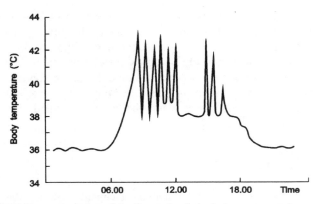

Fig. 18.1. Schematic representation of the fluctuation in body temperature of an antelope ground squirrel as it shuttles between the cool burrow and the hot surface for foraging. (After [22], with permission)

Maintaining resting T_{core} just above ambient temperature should be difficult for a non–sweating mammal, whose HP is several times larger than that of a similar–sized poikilotherm. However, heat dissipation is facilitated by a **greatly increased conductance**: the subcutaneous fat layer is thin, the mole rat is naked, and the fine and porous skin offers minimum resistance to the diffusion of water. Thus, passive evaporative heat loss helps to limit upward deviations of T_{core}.

The consequence of the high and essentially uncontrolled conductance is, of course, that the animal's ability to defend body temperature against external cold is compromised: outside and within the normally encountered environment, T_{core} of single naked mole rats is a linear function of ambient temperature. In fact, the relationship between both temperatures is indistinguishable from that in poikilotherms (Fig. 18.2, bottom).

However, the clear metabolic cold defence response at the lower end of the normally encountered range of ambient temperature (Fig. 18.2, top) classifies the species as a true homeotherm. As other rodents, it possesses brown adipose tissue, and noradrenaline induces non–shivering thermogenesis [223]. What is unique to mole rat is that even doubling HP at 29 °C ambient temperature is futile in the sense that it does not prevent T_{core} of a single individual from falling. The reason is the virtually complete lack of insulation which, on the other hand, is the prerequisite for regulating T_{core} close to ambient temperature. However, mole rats are highly social animals, and huddling behaviour attenuates the problems with the cold: a group of eight animals can stabilize T_{core} over an ambient temperature range of 4 °C [538]. Below this range, T_{core} and HP decline – a situation unlikely to be experienced except in a physiological laboratory.

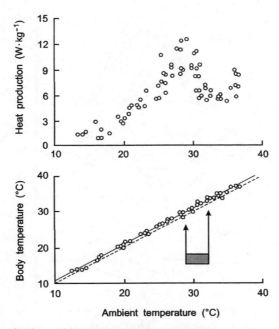

Fig. 18.2. Heat production and internal body temperature of single naked mole rats plotted vs. ambient temperature. *Grey bar and arrows* Normal range of ambient temperature in burrows; *interrupted line* identity of ambient and body temperatures. (After [70], with permission)

A general lesson to be learned from mole rat is that the usual 36–39 °C range of unstressed T_{core} in mammals is no constitutive feature of homeothermy. The normal and regulated body temperature range of a species could rather be seen as the consequence of optimizing resting HP and conductance in its habitat [312].

18.2
Larger Animals

Life in burrows is obviously no solution, and if the habitat is devoid of shade, **fur** becomes important **as a barrier against solar radiation**. The physical properties of fur have been dealt with in Chapter 7 – here, it is to document its efficiency as a means of preventing T_{core} from rising to levels which would require high rates of evaporative cooling.

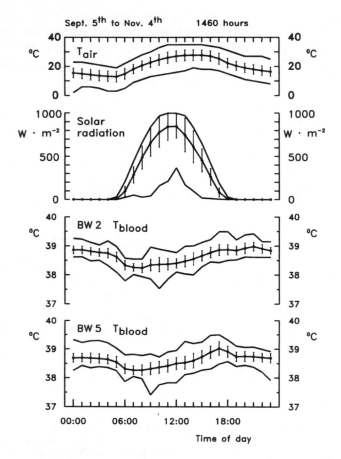

Fig. 18.3. Internal body temperature (*T*blood) of two free–ranging black wildebeest (*BW2*, *BW5*), air temperature (*T*air) and solar radiation plotted vs. time of the day. Means, standard deviations, minima and maxima of 1–h means of 61 days. (After [267], with permission)

In free–ranging black wildebeest, T_{core} was continuously measured over a period of 2 months. The dominant environmental factor was the intense solar radiation of up to 1000 W·m^{-2} at noon. The means and standard deviations of T_{core}, however, were 38.65 ±0.29 °C and 38.61 ±0.27 °C, respectively. The extremes averaged over an hour were 37.5 and 39.5 °C. In other words, the range of the animals' T_{core} in a very adverse radiation environment was indistinguishable from that of similar–sized animals in a temperate climate (Fig. 18.3).

The keys to this extraordinary performance are orientating behaviour and insulation by fur. Calculations based on measurements in a related species suggest that about 90% of the potential radiant heat load is prevented from flowing into the animals (Fig. 7.3). This leaves just 100 W·m^{-2}, plus 80 W·m^{-2} from resting HP, to be dissipated by evaporation. Wildebeest do not sweat, and the bulk of cooling has to be provided by panting, whose capacity for heat dissipation is modest in artiodactyls. Thus, fur and behaviour enable a species with limited means of evaporative cooling to lead at least a sedentary life in a habitat of intense solar radiation.

Camels go one step further. They have an unusual tolerance to dehydration: a resting animal endured 17 days in the Sahara without drinking water, leading to a 30% loss of its initial body mass [444,445]. However, the water loss was small in view of the duration of the dehydration period and the summer climate. One cause was that camels have perfected the mechanism of recuperation of water in the nose: in severe dehydration, the exhaled air is not just cooled but also desaturated so that it leaves the body at a relative humidity as low as 75%. The process of desaturation appears to be linked to the hygroscopic properties of dried and salty mucous in the nose, taking up water from the exhaled air and giving it off again during inhalation [446,451].

The second factor was the inhibition of sweating so that T_{core} during daytime rose to 40 °C or more, and the third was that T_{core} in the small hours fell to 35 °C or less. Both resulted in a largely augmented 24–h amplitude of T_{core}: up to 6 °C in dehydration vs. 2 °C in euhydration (Fig. 18.4). Commencing a hot day from a base line of 35 °C offers a distinct advantage to **animals of large body mass**. The balance, after 10 h and a rise in T_{core} to 40 °C, shows that a substantial fraction of the **heat load** arising from radiation and heat production (R and HP) **was stored temporarily** in the body. It resulted in a considerable saving of water: evaporative heat loss in dehydration was just 30% of that in euhydration.

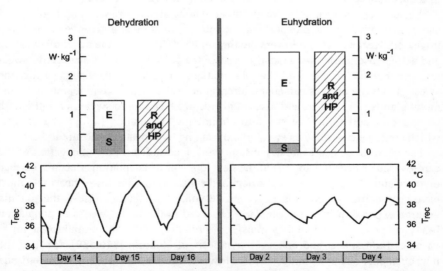

Fig. 18.4. 24–h cycle of a camel's body temperature in the desert during the last 3 days of a dehydration period (*left*), and euhydration (*right*). The *columns at the top* show a tentative heat balance for the 10 hottest hours of the day. *E* Evaporative heat loss; *S* heat storage; *R* radiant heat gain; *HP* heat production. (After [445], with permission)

There is another aspect. The heat load (indicated by the right columns in both panels) in dehydration amounted to only one half of that in euhydration. This was partly due to a significant reduction in HP [440,451]. The other reason was that, with progressing dehydration, the animal tended to expose as small an area as possible of the body surface to the sun, by assuming a sitting position with the legs under the body and the trunk orientated lengthwise to the direction of the sun's rays. As an addendum to the foregoing section, it is worth mentioning that the rate of water loss nearly doubled after shearing off the wool had increased the effective radiant heat load [445].

Augmented heat storage during daytime has also been observed in dehydrated oryx [490], and in euhydrated giraffe [301]. To large animals, it offers a means of coping with an environment in which hot days alternate with cool nights, so that the stored heat can be unloaded by convection and radiation, that is, without using water. It is also termed adaptive heterothermia and its decisive feature is a **wide interthreshold range** of T_{core} (Fig. 12.4): the elevated threshold temperature of sweating is combined with a lowered threshold of shivering.

18.3
Humans

The ability of humans to adapt to heat contrasts favourably with their limited capacity to adjust to repeated cold exposure. Regular exercise in a hot environment leads to a **higher sweat rate**. Improved evaporative cooling results in **lower skin and core temperatures**, and is accompanied by **lower heart rate** (Fig. 18.5).

A rise in T_{core} is the most important activating signal for sweating. Figure 18.5 shows that, after repeated exposures, the rate of sweating at a given level of T_{core} increased. Heat adaptation lowers the thresholds of T_{core} for the onset of sweating and active vasodilation [412], and increases the gain, that is, the increase in sweat rate per unit rise in T_{core} [376]. Observations in patas monkey suggest that the sweat glands themselves can adapt: after months of exposure to moderate heat, the sweat glands were larger, and excised glands produced more sweat per unit length of secretory coil following standard cholinergic stimulation [432] – a process which could be described as sweat gland training by repeated exposure to heat.

The decrease in heart rate at constant work load indicates the attenuation of the cardiovascular drift (Chap. 15) in the development of adaptation. Because cardiac output cannot be assumed to have declined substantially, the lower heart rate must have been compensated by a larger stroke volume. A likely cause is the **plasma volume expansion** which is regularly observed in the first 2 weeks of adaptation [434] and possibly mediated by transfer of protein from the interstitial to the intravascular fluid space, and enhanced synthesis of albumin [316,539]. Most of the improvements occur within 2 weeks. However, heat **adaptation is transient** and disappears gradually if it is not maintained by repeated exposure. Note the effects of weekends in Fig. 18.5: on Mondays, the subject regularly performed less efficiently than on the preceding Fridays.

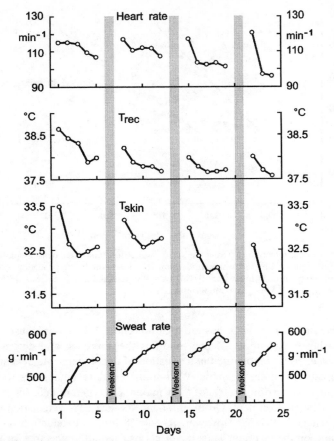

Fig. 18.5. Adaptation to exercise in the heat. At 45 °C air temperature and 25% relative humidity, a subject walked at 3 km·h^{-1} 3 h a day. The procedure was repeated daily for 24 days, except on weekends. *Sweat rate* is the mean over the 3–h period, T_{skin} (at the back), core temperature (T_{rec}) and heart rate are final values at the ends of exposure. (After [529], with permission)

Another prominent feature of heat adaptation is the production of **more dilute sweat**. At low sweat rates, the sodium concentration may be as low as 10 mEq·l^{-1}, resulting in a smaller loss of electrolytes and alleviating the side effects of sweating on plasma volume (Fig. 15.12). The electrolyte concentration of sweat does not entirely depend on the state of adaptation. Another important factor is sweat rate. Figure 18.6 shows that, at a given sweat rate, the concentration of sodium was considerably reduced after 3 weeks of adaptation. However, the enhanced capacity for sweating tends to counteract the effect of adaptation on sodium concentration. Apparently, the reuptake of sodium from the isotonic precursor fluid could not keep pace with increased secretion. Thus, if the enhanced capacity for sweating is used to work harder, the loss of electrolytes may still be substantial.

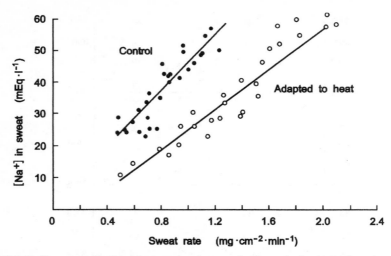

Fig. 18.6. Sodium concentration in sweat vs. sweat rate before and after heat adaptation. Local sweat rate was manipulated by hot water immersion. The heat adaptation procedure consisted in 3 weeks of daily exposure during which T$_{core}$ was maintained for 1 h at 38.0–38.5 °C. Data from one human subject. (After [5], with permission)

In the experiments of Fig. 18.5, heat adaptation reduced the strain caused by mild exercise in a hot environment. If one modifies the approach and determines the time strenuous exercise is tolerated in a situation of uncompensable heat stress, the effect of adaptation is more dramatic. Well–trained endurance athletes were asked to exercise until exhaustion at 60% of their maximal oxygen uptake for 9–12 days in a **hot and dry** environment. On the first and the last day, exhaustion occurred when T$_{core}$ reached 39.7 °C. However, it took 48 min to exhaustion on the first day, but 80 min on the last day. Thus the effect of adaptation was to nearly double **endurance time** [376].

Similar experiments in a **hot and humid** environment yielded even larger sweat rates than adaptation to dry heat, but most of the sweat dripped off, due to the physical limits to evaporation [378]. Adaptation to exercise in humid heat induces a more even distribution of sweating over the body surface. As a result, the **skin wettedness** (the fraction of the total body surface area which is covered by sweat) at a given elevated body temperature increases at the expense of some wasteful overproduction of sweat [340]. However, the higher sweat rate at the limbs promotes heat loss, owing to the large surfaces with particularly favourable conditions for evaporation [238].

19 Pathophysiology of Temperature Regulation

19.1
Hyperthermia and Heat Stroke

Internal body temperature (T_{core}) is bound to rise beyond the upper end of the tolerable range if heat production and external heat load continue to exceed heat loss. However, an intervening factor of major importance is what may be vaguely described as general physical fitness which varies between different subjects (heat–adapted vs. non–adapted, young vs. old, healthy vs. chronically ill), and can even change acutely in an individual (euhydrated vs. dehydrated). Thus the just tolerable combination of heat production and environment is subject to considerable variation, and so is the level of T_{core} at which pathological symptoms first occur: a supposedly fit helicopter pilot showed definite signs of developing heatstroke at 39.1 °C T_{core} [343], while the marathon runner of Fig. 15.5 succeeded in finishing after 160 min, despite T_{core} oscillating between 41.6 and 41.9 °C during the last 20 min of the race.

In humans, the signs of **heat exhaustion** comprise elevated T_{core}, weakness, fatigue, headache, nausea and diarrhoea. Heat exhaustion can quickly deteriorate to **classical heatstroke** whose symptoms are disturbance of central nervous functions (poor limb coordination, delirium, deep coma, convulsions), usually hot and dry skin owing to the absence of sweating, and high T_{core}. In the heatstroke cases during the 1982 Mekkah pilgrimage, T_{core} on arrival at the hospital was above 42 °C in 129 out of 178 patients [2]. Even with modern therapy, mortality is of the order of 10% [288]. Usually, the victims are older in age and have chronic illnesses, which may explain why mortality does not correlate closely with temperature on arrival [125]. Indeed, one patient 52 years of age and not having any relevant medical history, survived 46.5 °C T_{core} without permanent residua [472]. Classical heatstroke occurs often in epidemic form during prolonged heat waves. This and the absence of sweating suggest that **dehydration**, slowly developing during extended periods of sweating, could be pivotal in the development towards the final incident. Physical activity prior to classical heatstroke is typically described as low to moderate.

Exertional heatstroke, in contrast, occurs mainly in healthy young adults; the incidence is relatively high in motivated military recruits and competitive athletes. Sweating is often present: in fact, profuse sweating at the onset of heatstroke was reported in severe and finally fatal cases [456]. Mortality depends critically on the time elapsing before treatment commences; it was zero in 252 military recruits treated within 20 min after collapsing [97]. The important conclusion is that the **combination of temperature and exposure time** predicts the outcome and chance of survival better than temperature alone [74,240].

Clinical observations in very ill patients are not very suitable for elucidating the pathogenesis of heatstroke, for the simple reason that reconstructing the conditions conducive to the incident is mostly impossible. Thus, the pathogenetic concepts are based on long extrapolations from responses of healthy human subjects to innocuous degrees of exercise hyperthermia, and on animal experiments.

As was detailed in Chapters 10 and 15, the **cardiovascular state** during severe exercise or in a hot environment is characterized by high skin blood flow. With progressive hyperthermia, central blood volume and venous pressure decline, owing to the combined effects of blood volume translocation to the skin, ortho-static pressure and dehydration. While extremely fit endurance athletes were able to maintain adequate levels of cardiac output even at the point of exhaustion at 39.9 °C T_{core} [378], it appears possible that, in less adapted people at the stage of imminent heatstroke, decreasing **central venous pressure** reduces cardiac output below a critical minimum [165]. The results of animal experiments were not al-ways definite [172,173].

A regular occurrence in hyperthermia is **splanchnic vasoconstriction**, partly compensating for skin vasodilation (Fig. 15.8). Its degree depends on the relation-ship between fitness and severity of thermal or exercise stress [426]. It is conceiv-able that intense and prolonged ischemia is detrimental to the barrier function of the gut. Its lumen contains large amounts of toxic **lipopolysaccharides** (LPS) which are shed from the walls of gut bacteria. In anaesthetized and heat–stressed monkeys, the plasma concentration of LPS rose when T_{core} exceeded 42 °C, and the increase occurred first in the portal vein, suggesting the intestinal origin of LPS (Fig. 19.1).

Above 39 and below 42 °C, the concentrations of **natural antibodies against LPS** (immunoglobulins class G) was clearly higher in the portal vein than in the femoral artery. This suggests that LPS entered the circulation already within this temperature range; however, they bound to anti–LPS and were cleared, by the liver, from the blood so that the systemic concentration of LPS downstream from the liver remained low. Only in the last phase prior to heatstroke did the concen-tration become too high to be bound completely, and the normally low rate of overspill of LPS into the general circulation increased. As in fever, higher plasma concentrations of LPS induce the release of **cytokines** (Chap. 16) which mediate many symptoms of septic shock, a clinical syndrom sharing multiple–organ failure with heatstroke [49].

Support for the hypothesis of an important role of gut LPS comes from two sides. First, reducing the gut flora by non–resorbable antibiotics or lavage in ex-perimental animals attenuated the rise in body temperature during heat exposure and increased the rate of survival in heatstroke [72,73]. Second, the concentrations of LPS and cytokines in the plasma of heatstroke victims were greatly elevated on admission to the hospital, and declined after cooling [49].

The possible link between **heatstroke and fever** causes a conceptual conflict. Uncontrolled hyperthermia in healthy subjects, leading potentially to heatstroke, is thought to occur in spite of the heat–dissipating mechanisms operating at full capacity. Fever, in contrast, implies an elevation of the set–point, causing high T_{core} to be defended by inhibition of heat dissipation and/or activation of heat production. If, then, fever is part of heatstroke, one would expect to see symptoms indicative of a higher set–point. The evidence is equivocal. Classical heatstroke is typically associated with dry skin, but exertional heatstroke with sweating, although not necessarily of high intensity. Occasionally, heatstroke victims undergoing rigorous cooling procedures were reported to show goose–flesh skin and shivering at still elevated T_{core} [72,73]. On balance, the distinction between hyperthermia and fever appears to lose significance at the stage of heatstroke [174].

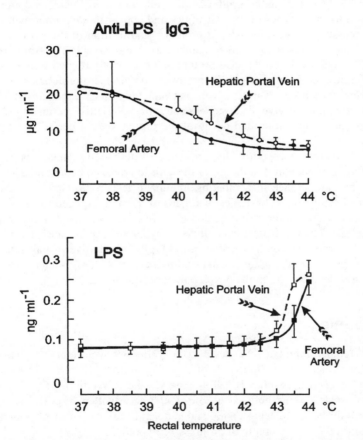

Fig. 19.1. Plasma anti–LPS antibodies (*Anti–LPS IgG, top*) and lipopolysaccharides (*LPS, bottom*) vs. body temperature in monkeys. LPS and Anti–LPS IgG were sampled in the portal vein collecting the venous return from the intestines, and in a femoral artery, downstream from the liver. (After [140], with permission)

A large number of studies investigated the effects of severe heat exposure at the **cellular level**, mostly on cell cultures [169]. A regular finding, apart from structural damage to the cytoskeleton, was that the rates of passive flux and active transport of ions across cell membranes increased in parallel with rising temperature. One model holds that the energy cost of active transport rises out of proportion to the rate of transport, ultimately leading to energy depletion and subsequent fatal disturbance of the ionic milieu of the cells [239]. Protein synthesis is inhibited at higher temperature, and increased cellular injury is documented by excessively high extracellular levels of marker enzymes. However, no single critical factor has been identified so far, and there may be none. Rather, a multitude of effects may interact and limit the tolerance time at high temperature.

More recently, interest has focused on the role of **heat shock proteins (HSPs)**. HSPs from 27–110 kDa can be synthesized by all organisms, plants and animals, as a response to many types of cellular injury and appear to have important functions in processing of stress–denatured proteins, stabilization of the cytoskeleton, and maintenance of endothelial and epithelial barrier functions [527]. Amongst the various HSPs, the level of a 72–kDa protein is most closely related to the magnitude and duration of thermal stress. Its presence confers thermotolerance: a cellular adaptation caused by a single, severe but non–lethal heat exposure which allows the organism to survive a subsequent and otherwise lethal thermal stress and is linked to the accumulation and decay of the protein over a period of hours to days [353].

Outside a laboratory, repetitive hyperthermia of potentially dangerous degrees is not seen very often, and it is therefore difficult to envisage beneficial effects of HSPs and thermotolerance in combating imminent heatstroke. However, the accumulation of HSPs in single tissues or organs of animals exposed to severe heat stress can serve as a marker of injury, and identify certain organs as more susceptible to thermal damage than others. Indeed, hyperthermia in passively heated, non–exercising rats induced large increases in HSP72 in the liver, small intestine, and kidneys, but not in the brain and skeletal muscle [128]. This points once more to the critical role of the gut in the development of heatstroke.

19.2
Hypothermia

At low skin temperature, a 2–3 °C deviation of T_{core} below normal is sufficient to elicit maximum heat production (HP) of a resting animal [189,366]. If heat loss continues to exceed maximum HP, T_{core} passes the lower limit of the zone of proportional control. This defines the **onset of uncontrolled hypothermia**. HP declines with further decreasing T_{core}, and **general deceleration of physiological mechanisms** is the key symptom of progressive hypothermia [389].

The dependence of maximum HP on T_{core} is shown in Fig. 19.2. Shorn sheep were wetted with cold water and exposed to a wind of 7 m·s^{-1} at 0 °C. At 30 °C T_{core}, HP was just 30% of the maximum at 38 °C [27]. Extrapolation predicts that, with further falling temperature, metabolic cold defence would have completely ceased near 27 °C; below this level, actual HP would equal basal metabolic rate at this temperature and decline according to the general temperature dependence of biological reactions (Arrhenius–van't Hoff effect). The reduced metabolic rate at low temperature increases the survival time at circulatory arrest and is the rationale behind the use of artificial hypothermia in surgery [470,471,489].

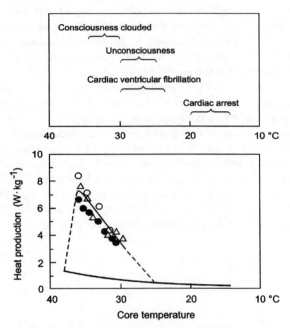

Fig. 19.2. General symptoms of progressive hypothermia (*top*), and maximum heat production (*bottom*) plotted vs. body temperature. *Single symbols* show data from experiments in three conscious sheep. *Interrupted lines* are extrapolated. The *heavy solid line* shows basal heat production, assuming a Q_{10} of 2.5. (Sheep data from [27], with permission)

The older literature contains a number of reports on circulation and respiration in deep hypothermia. Because of the interfering effects of anaesthesia, interpretation of the studies is difficult. On balance, however, it appears that, once the range of metabolic cold defence is transgressed or bypassed by anaesthesia, gross functional parameters such as respiratory minute ventilation and cardiac output decline in parallel with metabolic rate [501]. Thus, even at rather low temperatures, the oxygen supply to the tissues remains adequate in view of the reduced consumption. The prerequisite is, of course, that the **heart** keeps beating. Particularly in the temperature range between 30 and 25 °C, the heart is susceptible to potentially lethal ventricular fibrillation, the causes of which are still unknown [389]. If this range is passed, circulation is maintained until cardiac arrest occurs eventually at about 15 °C. Nothing of the kind happens in hibernators or the very young of altricial species (Chap. 17). A possibly important difference concerns the adrenergic innervation of the heart. In hibernators, it is confined to the conduction system and the coronary vessels, and leaves out the ventricular myocytes; a factor supposed to reduce the calcium overload during hypothermia [270].

Hibernators or torpor species may even survive short periods of what one would consider clinical death in other circumstances. In experiments on tree bats exposed to −10 °C T_{air}, two animals completely stopped breathing. The bats were found on the floor of the chamber and thought to be dead. T_{core} could not be measured since the body was too hard to permit introduction of a temperature probe. After a few minutes at 23 °C T_{air}, however, the bats slowly resumed breathing, and flew again in the evening [144].

In humans, **higher central nervous functions** show first signs of impairment at relatively small decreases in T_{core} [389]. Near 34 °C, memory registration and mental performance of more complex tasks deteriorated [93,146]. With progressive hypothermia, slurring of speech and disorientation developed. Most individuals lose consciousness at 28–30 °C [395]. Fig. 19.3 shows the relationship between seawater temperature and immersion time in shipwreck survivors picked up from the ocean by the United States Navy in World War II. The line of survival falls steeply at water temperatures near 20 °C, indicating that hypothermia became increasingly important as the cause of drowning. At water temperatures below 5 °C survival time was a matter of minutes rather than of hours.

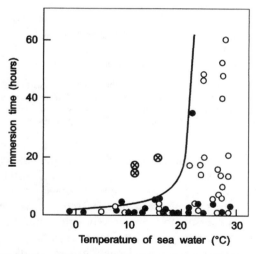

Fig.19.3. Duration of immersion of shipwreck survivors plotted vs. seawater temperature. A single *symbol* indicates survival of at least one man; *closed circles* water temperature taken at time and location of rescue; *open circles* temperature taken from a world atlas of sea–surface temperatures; *crosses in circles* the slowest successful participant in the 1951 swimming race across the English Channel, and two swimmers finishing the 1953 Nile marathon. (Shipwreck data from [347], swimmer data from [396], with permission)

Some people show outstanding endurance in cold water. In the 1950s (before the introduction of wet suits), swimming races over long distances, such as across the English Channel from France to England, were popular. This was a remarkable endeavour: it required the ability to swim in water of approximately 15 °C for periods between 12 and 20 h, which is far outside the line of survival in Fig. 19.3. Those who succeeded had two features in common that are not often seen together in human athletes: endurance to sustain high levels of exercise for many hours, and obesity. In fact, all swimmers were fat, and many of them grossly fat. The internal conductance (Chap. 7) was very low, and T_{core} of one subject immediately after the race was as high as 36.7 °C [396]. Thus, in certain situations, being fat apparently confers advantages not only on seals and whales, but also humans.

However, this does not imply that large body mass and plenty of subcutaneous fat are indispensable for avoiding hypothermia in cold water. A rather unique example is the semiaquatic **platypus**, one of the few extant egg–laying mammals whose ancestors separated from the line leading to present–day placentals at least 100 million years ago. It is living near the lakes and rivers of eastern Australia, down to Tasmania in the south, and is a so–called primitive mammal: minimum HP during rest is 35% lower than in advanced mammals of equal body mass (1–2 kg), and unstressed T_{core} is 31–32 °C [112]. Nevertheless, the platypus is an excellent homeotherm. A free–ranging animal spent more than 10 h in a river of 5 °C [155]. T_{core} in water was essentially the same as in air (Fig. 19.4).

Platypus is nearly devoid of subcutaneous fat. However, the water–repellent fur on the squat body consists in a thick mass of short woolly underfur, overlain by guard hairs. More than 800 fibres mm^{-2} were found on the dorsal surface. In the unfurred limbs, the arteries are paralleled by one or more venous vessels, and the arrangement is suggestive of a sophisticated countercurrent heat–exchange system [156]. It is mainly the combination of fur and vascular tissue insulation that enables this small species to maintain a temperature gradient of nearly 30 °C between body core and water; the metabolic rate at 5 °C was only 1.5 times larger than that at water temperatures of 20–30 °C. Apparently, taxonomic primitiveness does not exclude perfect homeothermy in an adverse environment.

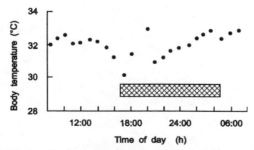

Fig. 19.4. Body temperature of a free-ranging platypus in air, and in water of 5 °C. Body temperature was measured by radiotelemetry. *Each point* represents the mean of four readings taken at 15-min intervals; *cross-hatched bar* animal in water. (After [155], with permission)

References

1. Adachi A (1984) Thermosensitive and osmoreceptive afferent fibers in the hepatic branch of the vagus nerve. J Auton Nerv Syst 10: 269–273
2. Al–Khawashki MI, Mustafa MKY, Khogali M, El–Sayed H (1983) Clinical presentation of 172 heat stroke cases seen at Mina and Arafat–September, 1982. In: Khogali M, Hales JRS (eds) Heat stroke and temperature regulation. Academic Press, Sydney, pp 99–108
3. Alexander G, Bell AW, Hales JRS (1973) Effects of cold exposure on tissue blood flow in the new–born lamb. J Physiol (Lond) 234: 65–77
4. Alexander G, Williams D (1968) Shivering and non–shivering thermogenesis during summit metabolism in young lambs. J Physiol (Lond) 198: 251–276
5. Allan JR, Wilson CG (1971) Influence of acclimatization on sweat sodium concentration. J Appl Physiol 30: 708–712
6. Amini–Sereshki L, Zarrindast MR (1984) Brain stem tonic inhibition of thermoregulation in the rat. Am J Physiol 247: R154–R159
7. Andersen P, Saltin B (1985) Maximal perfusion of skeletal muscle in man. J Physiol (Lond) 366: 233–249
8. Andersson B, Gale CC, Hoekfelt B, Larsson B (1965) Acute and chronic effects of pre-optic lesions. Acta Physiol Scand 65: 45–60
9. Arman CG,van, Armstrong DAJ, Kim DH (1991) Antipyretics. In: Schönbaum E, Lomax P (eds) Thermoregulation: pathology, pharmacology, and therapy. Pergamon Press, New York, pp 55–104
10. Aschoff J (1981) Thermal conductance in mammals and birds: its dependence on body size and circadian phase. Comp Biochem Physiol (A) 69: 611–619
11. Aschoff J, Wever R (1958) Kern und Schale im Wärmehaushalt des Menschen. Naturwissenschaften 45: 477–485
12. Baker MA (1981) Anatomical and physiological adaptations of panting animals to heat and exercise. In: Horvath SM, Yousef MK (eds) Environmental physiology: aging, heat and altitude. Elsevier North–Holland, New York, pp 121–146
13. Baker MA (1982) Brain cooling in endotherms in heat and exercise. Annu Rev Physiol 44: 85–96
14. Baker MA (1989) Effects of dehydration and rehydration on thermoregulatory sweating in goats. J Physiol (Lond) 417: 421–435
15. Baker MA, Chapman LW (1977) Rapid brain cooling in exercising dogs. Science 195: 781–783
16. Baker MA, Turlejska E (1989) Thermal panting in dehydrated dogs: effects of plasma volume expansion and drinking. Pflügers Arch 413: 511–515
17. Baldwin BA, Yates JO (1977) The effects of hypothalamic temperature variation and intra-carotid cooling on behavioural thermoregulation in sheep. J Physiol (Lond) 265: 705–720
18. Banet M, Hensel H, Liebermann H (1978) The central control of shivering and non–shivering thermogenesis in the rat. J Physiol (Lond) 283: 569–584
19. Barbour HG (1912) Die Wirkung unmittelbarer Erwärmung und Abkühlung der Wärmezentra auf die Körpertemperatur. Arch Exp Pathol Pharmakol 70: 1–15
20. Barker JL, Carpenter DO (1970) Thermosensitivity of neurons in the sensorimotor cortex of the cat. Science 169: 597–598
21. Barnes BM (1989) Freeze avoidance in a mammal: body temperatures below 0°C in an arctic hibernator. Science 244: 1593–1595
22. Bartholomew GA (1964) The roles of physiology and behaviour in the maintenance of homeostasis in the desert environment. In: Hughes GM (ed) Homeostasis and feedback

mechanisms. Symposia of the Society for Experimental Biology, vol XVIII. Cambridge University Press, Cambridge, pp 7–29

23. Bawa P, Matthews PBC, Mekjavic IB (1987) Electromyographic activity during shivering of muscles acting at the human elbow. J Therm Biol 12: 1–4
24. Bazett HC (1927) Physiological responses to heat. Physiol Rev 7: 531–599
25. Bazett HC, Love L, Newton M, Eisenberg L, Day R, Forster R,II (1948) Temperature changes in blood flowing in arteries and veins in man. J Appl Physiol 1: 3–19
26. Beaumont W,van, Bullard RW (1963) Sweating: its rapid response to muscular work. Science 141: 643–646
27. Bennett JW (1972) The maximum metabolic response of sheep to cold: Effects of rectal temperature, shearing, feed consumption, body posture, and body weight. Aust J Agric Res 23: 1045–1058
28. Bennett NC, Jarvis JUM, Davies KC (1988) Daily and seasonal temperatures in the burrows of African rodent moles. S Afr J Zool 23: 189–195
29. Benzinger TH (1969) Heat regulation: homeostasis of central temperature in man. Physiol Rev 49: 671–759
30. Benzinger TH, Taylor GW (1963) Cranial measurements of internal temperature in man. In: Hardy JD (ed) Temperature: its measurement and control in science and industry, vol III/3. Reinhold Publishing Corporation, New York, pp 111–120
31. Bergersen TK, Eriksen M, Wallöe L (1995) Effect of local warming on hand and finger artery blood velocities. Am J Physiol 269: R325–R330
32. Bergersen TK, Eriksen M, Wallöe L (1997) Local constriction of arteriovenous anastomoses in the cooled finger. Am J Physiol 273: R880–R886
33. Bergersen TK, Hisdal J, Wallöe L (1999) Perfusion of human finger during cold–induced vasodilatation. Am J Physiol 276: R731–R737
34. Beshir MY, Ramsey JD (1988) Heat stress indices: a review paper. Int J Ind Ergonom 3: 89–102
35. Bianca W (1965) Sweating in dehydrated steers. Res Vet Sci 6: 33–37
36. Bignall KE, Schramm L (1974) Behavior of chronically decerebrated kittens. Exp Neurol 42: 519–531
37. Bijman I, Quinton PM (1984) Predominantly beta–adrenergic control of equine sweating. Am J Physiol 246: R349–R353
38. Blatteis CM, Sehic E (1997) Fever: how may circulating pyrogens signal the brain? News Physiol Sci 12: 1–9
39. Blatteis CM, Sehic E (1997) Prostaglandin E2: a putative fever mediator. In: Mackowiak PA (ed) Fever: basic mechanisms and management. Lippincott–Raven, Philadelphia, PA, pp 117–145
40. Blatteis CM, Sehic E, Li S (1998) Afferent pathways of pyrogen signaling. In: Kluger MJ, Bartfei T, Dinarello C (eds) Molecular mechanisms of fever (Annals of the New York Academy of Sciences, vol 856). New York Academy of Sciences, New York, pp 95–107
41. Blaxter KL (1989) Energy metabolism in animals and man. Cambridge University Press, Cambridge, pp 1–336
42. Bleichert A, Behling K, Kitzing J, Scarperi M, Scarperi S (1972) Antriebe und effektorische Maßnahmen der Thermoregulation bei Ruhe und während körperlicher Arbeit. IV. Ein analoges Modell der Thermoregulation bei Ruhe und Arbeit. Int Z Angew Physiol 30: 193–206
43. Bleichert A, Behling K, Scarperi M, Scarperi S (1973) Thermoregulatory behavior of man during rest and exercise. Pflügers Arch 338: 303–312
44. Bligh J (1957) The relationship between the temperature in the rectum and of the blood in the bicarotid trunk of the calf during exposure to heat stress. J Physiol (Lond) 136: 393–403
45. Bligh J (1966) The thermosensitivity of the hypothalamus and thermoregulation in mammals. Biol Rev Camb Philos Soc 41: 317–367
46. Bligh J (1972) Neuronal models of mammalian temperature regulation. In: Bligh J, Moore RE (eds) Essays on temperature regulation. North–Holland, Amsterdam, pp 105–120
47. Bligh J (1973) Temperature regulation in mammals and other vertebrates. North–Holland, Amsterdam, pp 1–436

48. Blix AS, Steen JB (1979) Temperature regulation in newborn polar homeotherms. Physiol Rev 59: 285–304

49. Bouchama A, Parhar RS, El–Yazigi A, Sheth K, Al–Sedairy S (1991) Endotoxemia and release of tumor necrosis factor and interleukin 1α in acute heatstroke. J Appl Physiol 70: 2640–2644

50. Boulant JA (1980) Hypothalamic control of thermoregulation: neurophysiological basis. In: Morgane PJ, Panksepp J (eds) Handbook of the hypothalamus, vol III, part A: Behavioral studies. Dekker, New York, pp 1–82

51. Boulant JA (1994) Neurophysiology of thermoregulation: role of hypothalamic neuronal networks. In: Milton AS (ed) Temperature regulation: recent physiological and Pharmacological advances. Birkhäuser, Basel, pp 93–101

52. Boulant JA (1996) Hypothalamic neurons regulating body temperature. In: Fregly MJ, Blatteis CM (eds) Handbook of physiology, sect 4: Environmental physiology, vol I. Oxford University Press, New York, pp 105–126

53. Boulant JA, Dean JB (1986) Temperature receptors in the central nervous system. Annu Rev Physiol 48: 639–654

54. Boulant JA, Hardy JD (1974) The effect of spinal and skin temperatures on the firing rate and thermosensitivity of preoptic neurons. J Physiol (Lond) 240: 639–660

55. Boyer BB, Barnes BM, Lowell BB, Grujic D (1998) Differential regulation of uncoupling proteine gene homologues in multiple tissues of hibernating ground squirrels. Am J Physiol 275: R1232–R1238

56. Böckler H, Heldmaier G (1983) Interaction of shivering and non–shivering thermogenesis in seasonally–acclimatized Djungarian hamsters (*Phodopus sungorus*). J Therm Biol 8: 97–98

57. Bradley SR, Deavers DR (1980) A re–examination of the relationship between thermal conductance and body weight in mammals. Comp Biochem Physiol (A) 65: 465–476

58. Braun HA, Schäfer K, Wissing H (1990) Theories and models of temperature transduction. In: Bligh J, Voigt K (eds) Thermoreception and temperature regulation. Springer, Berlin, pp 19–29

59. Brengelmann GL (1983) Circulatory adjustments to exercise and heat stress. Annu Rev Physiol 45: 191–212

60. Brengelmann GL (1987) Dilemma of body temperature measurement. In: Shiraki K, Yousef MK (eds) Man in stressful environments – thermal and work physiology. Thomas, Springfield, pp 5–22

61. Brengelmann GL (1993) Specialized brain cooling in humans? FASEB J 7: 1148–1153

62. Brengelmann GL, Freund PR, Rowell LB, Olerud JE, Kraning KK (1981) Absence of active cutaneous vasodilation associated with congenital absence of sweat glands in humans. Am J Physiol 240: H571–H575

63. Briese E (1986) Circadian body temperature rhythm and behavior of rats in thermoclines. Physiol Behav 37: 839–847

64. Brinnel H, Cabanac M (1989) Tympanic temperature is a core temperature in humans. J Therm Biol 14: 47–53

65. Brinnel H, Cabanac M, Hales JRS (1987) Critical upper limits of body temperature, tissue thermosensitivity and selective brain cooling in hyperthermia. In: Hales JRS, Richards DAB (eds) Heat stress: physical exertion and environment. Elsevier Science Publishers B.V., Amsterdam, pp 209–240

66. Brody S (1945) Bioenergetics and growth. With special reference to the efficiency complex in domestic animals. Reinhold, New York, pp 1–1023

67. Brown AC, Brengelmann GL (1970) The interaction of peripheral and central inputs in the temperature regulating system. In: Hardy JD, Gagge AP, Stolwijk JAJ (eds) Physiological and behavioral temperature regulation. Thomas, Springfield, pp 684–702

68. Brück K (1978) Heat production and temperature regulation. In: Stave U (ed) Perinatal Physiology. Plenum, New York, pp 455–498

69. Brück K, Wünnenberg W (1970) "Meshed" control of two effector systems: nonshivering and shivering thermogenesis. In: Hardy JD, Gagge AP, Stolwijk JAJ (eds) Physiological and behavioral temperature regulation. Thomas, Springfield, pp 562–580

70. Buffenstein R, Yahav S (1991) Is the naked mole–rat *Heterocephalus glaber* an endo-
     thermic yet poikilothermic mammal? J Therm Biol 16: 227–232
71. Burton AC, Edholm OG (1969) Man in a cold environment. Hafner Publishing Company,
     New York, pp 1–273
72. Butkow N, Mitchell D, Laburn HP, Kenedi E (1984) Heat stroke and endotoxaemia in
     rabbits. In: Hales JRS (ed) Thermal physiology. Raven Press, New York, pp 511–514
73. Bynum G, Brown J, Dubose D, Marsili M, Leav I, Pistole TG, Hamlet M, LeMaire M,
     Caleb B (1979) Increased survival in experimental dog heatstroke after reduction of gut
     flora. Aviat Space Environ Med 50: 816–819
74. Bynum GD, Pandolf KB, Schuette WH, Goldman RF, Lees DE, Whang–Peng J, Atkinson
     ER, Bull JM (1978) Induced hyperthermia in sedated humans and the concept of critical
     thermal maximum. Am J Physiol 235: R228–R236
75. Cabanac M (1993) Selective brain cooling in humans: "fancy" or fact? FASEB J 7: 1143–
     1147
76. Cabanac M (1996) Heat stress and behavior. In: Fregly MJ, Blatteis CM (eds) Handbook
     of physiology, sect 4: Environmental physiology, vol I. Oxford University Press, New
     York, pp 261–278
77. Cabanac M, Massonnet B, Belaiche R (1972) Preferred skin temperature as a function of
     internal and mean skin temperature. J Appl Physiol 33: 699–703
78. Calvert DT, Findlay JD, McLean JA (1981) Quantitative aspects of preoptic thermo-
     sensitivity in the conscious ox. Q J Exp Physiol 66: 377–390
79. Canals M, Rosenmann M, Bozinovic F (1989) Energetics and geometry of huddling in
     small mammals. J theor Biol 141: 181–189
80. Candas V, Libert JP, Vogt JJ (1983) Sweating and sweat decline of resting men in hot
     humid environments. Eur J Appl Physiol 50: 223–234
81. Caputa M, Feistkorn G, Jessen C (1986) Effects of brain and trunk temperatures on exer-
     cise performance in goats. Pflügers Arch 406: 184–189
82. Caputa M, Kadziela W, Narebski J (1976) Significance of cranial circulation for the brain
     homeothermia in rabbits. I. The brain–arterial blood temperature gradient. Acta Neurobiol
     Exp 36: 613–624
83. Caputa M, Kamari A, Wachulec M (1991) Selective brain cooling in rats resting in heat
     and during exercise. J Therm Biol 16: 19–24
84. Carlisle HJ, Ingram DL (1973) The effects of heating and cooling the spinal cord and
     hypothalamus on thermoregulatory behaviour in the pig. J Physiol (Lond) 231: 353–364
85. Carpenter DO (1981) Ionic and metabolic bases of neuronal thermosensitivity. Fed Proc
     40: 2808–2813
86. Castellani JW, Young AJ, Sawka MN, Pandolf KB (1998) Human thermoregulatory
     responses during serial cold–water immersions. J Appl Physiol 85: 204–209
87. Chambers WW, Seigel MS, Liu JC, Liu CN (1974) Thermoregulatory responses of de-
     cerebrate and spinal cats. Exp Neurol 42: 282–299
88. Chen XM, Hosono T, Yoda T, Fukuda Y, Kanosue K (1998) Efferent projection from the
     preoptic area for the control of non–shivering thermogenesis in rats. J Physiol (Lond) 512:
     883–892
89. Clapperton JL, Joyce JP, Blaxter KL (1965) Estimates of the contribution of solar radi-
     ation to the thermal exchanges of sheep at a latitude of 55° north. J Agric Sci (Camb) 64:
     37–49
90. Clark RP, Edholm OG (1985) Man and his thermal environment. Edward Arnold, London,
     pp 1–253
91. Clark WG, Fregly MJ (1996) Evidence for roles of brain peptides in thermoregulation. In:
     Fregly MJ, Blatteis CM (eds) Handbook of physiology, sect 4: Environmental physiology,
     vol I. Oxford University Press, New York, pp 139–153
92. Coceani F, Akarsu ES (1998) Prostaglandin E2 in the pathogenesis of fever. An update. In:
     Kluger MJ, Bartfei T, Dinarello C (eds) Molecular mechanisms of fever (Annals of the
     New York Academy of Sciences, vol 856). New York Academy of Sciences, New York,
     pp 76–82

93. Coleshaw SRK, Someren RNM,van, Wolff AH, Davis HM, Keatinge WR (1983) Impaired memory registration and speed of reasoning caused by low body temperature. J Appl Physiol 55: 27–31
94. Colquhoun EQ, Clark MG (1991) Open question: has thermogenesis in muscle been overlooked and misinterpreted? News Physiol Sci 6: 256–259
95. Cooper KE (1995) Fever and antipyresis. Cambridge University Press, Cambridge, pp 1–182
96. Cooper KE, Kenyon JR (1957) A comparison of temperatures measured in the rectum, oesophagus and on the surface of the aorta during hypothermia in man. Br J Surg 44: 616–619
97. Costrini A (1990) Emergency treatment of exertional heatstroke and comparison of whole body cooling techniques. Med Sci Sports Exerc 22: 15–18
98. Crawford EC, Jr. (1962) Mechanical aspects of panting in dogs. J Appl Physiol 17: 249–251
99. Crawshaw LI, Nadel ER, Stolwijk JAJ, Stamford BA (1975) Effect of local cooling on sweating rate and cold sensation. Pflügers Arch 354: 19–27
100. Cronin MJ, Baker MA (1977) Physiological responses to midbrain thermal stimulation in the cat. Brain Res 128: 542–546
101. Cronin MJ, Baker MA (1977) Thermosensitive midbrain units in the cat. Brain Res 128: 461–472
102. Cumming DHM (1975) A field study of the ecology and behaviour of warthog. Mus Mem Natl Mus Monum Rhod 7: 1–179
103. Daniel PM, Dawes JDK, Prichard MML (1953) Studies of the carotid rete and its associated arteries. Philos Trans R Soc Lond (Biol) 237: 173–237
104. Davies CTM (1979) Thermoregulation during exercise in relation to sex and age. Eur J Appl Physiol 42: 71–79
105. Davies CTM (1979) Influence of skin temperature on sweating and aerobic performance during severe work. J Appl Physiol 47: 770–777
106. Davies SN (1985) Sympathetic modulation of cold–receptive neurones in the trigeminal system of the rat. J Physiol (Lond) 366: 315–329
107. Davies SN, Goldsmith GE, Hellon RF, Mitchell D (1983) Facial sensitivity to rates of temperature change: Neurophysiological and psychophysical evidence from cats and humans. J Physiol (Lond) 344: 161–175
108. Davies SN, Goldsmith GE, Hellon RF, Mitchell D (1985) Sensory processing in a thermal afferent pathway. J Neurophysiol 53: 429–434
109. Davis KD, Lozano AM, Manduch M, Tasker RR, Kiss ZHT, Dostrovsky O (1999) Thalamic relay site for cold perception in humans. J Neurophysiol 81: 1970–1973
110. Dawes JDK, Prichard MML (1953) Studies of the vascular arrangements of the nose. J Anat 87: 311–326
111. Dawson TJ (1972) Primitive mammals and patterns in the evolution of thermoregulation. In: Bligh J, Moore RE (eds) Essays on temperature regulation. North–Holland, Amsterdam, pp 1–18
112. Dawson TJ (1989) Responses to cold of monotremes and marsupials. In: Wang LCH (ed) Advances in comparative and environmental physiology, vol 4. Springer, Berlin Heidelberg, pp 255–288
113. Dawson TJ, Hulbert AJ (1970) Standard metabolism, body temperature, and surface areas of Australian marsupials. Am J Physiol 218: 1233–1238
114. Dawson TJ, Robertshaw D, Taylor CR (1974) Sweating in the kangaroo: a cooling mechanism during exercise, but not in the heat. Am J Physiol 227: 494–498
115. Dickenson AH (1977) Specific responses of rat raphe neurones to skin temperature. J Physiol (Lond) 273: 277–293
116. Dmi'el R (1986) Selective sweat secretion and panting modulation in dehydrated goats. J Therm Biol 11: 157–160
117. Dmi'el R, Prevulotzky A, Shkolnik A (1980) Is a black coat in the desert a means of saving metabolic energy? Nature 283: 761–762
118. Dmi'el R, Robertshaw D (1983) The control of panting and sweating in the black Bedouin goat: a comparison of two modes of imposing a heat load. Physiol Zool 56: 404–411

119. Dubner R, Sumino R, Wood WJ (1975) A peripheral cold fiber population responsive to innocuous and noxious thermal stimuli applied to the monkey's face. J Neurophysiol 38: 1373–1389

120. Entin PL, Robertshaw D, Rawson RE (1998) Thermal drive contributes to hyperventilation during exercise in sheep. J Appl Physiol 85: 318–325

121. Evans DL, Rose RJ (1987) Cardiovascular and respiratory responses to submaximal exercise training in the thoroughbred horse. Pflügers Arch 411: 316–321

122. Farrell DM, Bishop VS (1995) Permissive role for nitric oxide in active thermoregulatory vasodilation in rabbit ear. Am J Physiol 269: H1613–H1618

123. Feistkorn G, Nagel A, Jessen C (1984) Circulation and acid–base balance in exercising goats at different body temperatures. J Appl Physiol 57: 1655–1661

124. Ferris BG,Jr., Forster RE,II, Pillion EL, Christensen WR (1947) Control of peripheral blood flow: responses in the human hand when extremities are warmed. Am J Physiol 150: 304–314

125. Ferris EB,Jr., Blankenhorn MA, Robinson HW, Cullen GE (1938) Heat stroke: clinical and chemical observations on 44 cases. J clin Invest 17: 249–262

126. Finch VA (1972) Energy exchanges with the environment of two East African antelopes, the eland and the hartebeest. Symp Zool Soc (Lond) 31: 315–326

127. Finch VA (1972) Thermoregulation and heat balance of the East African eland and hartebeest. Am J Physiol 222: 1374–1379

128. Flanagan SW, Ryan AJ, Gisolfi CV, Moseley PL (1995) Tissue–specific HSP70 response in animals undergoing heat stress. Am J Physiol 268: R28–R32

129. Flavahan NA (1991) The role of vascular $\alpha2$–adrenoceptors as cutaneous thermosensors. News Physiol Sci 6: 251–255

130. Florant GL, Turner BM, Heller HC (1978) Temperature regulation during wakefulness, sleep, and hibernation in marmots. Am J Physiol 235: R82–R88

131. Folkow LP, Mercer JB (1986) Partition of heat loss in resting and exercising winter– and summer–insulated reindeer. Am J Physiol 251: R32–R40

132. Fortney SM, Nadel ER, Wenger CB, Bove JR (1981) Effect of blood volume on sweating rate and body fluids in exercising humans. J Appl Physiol 51: 1594–1600

133. Fortney SM, Wenger CB, Bove JR, Nadel ER (1984) Effect of hyperosmolality on control of blood flow and sweating. J Appl Physiol 57: 1688–1695

134. Foster DO (1986) Quantitative role of brown adipose tissue in thermogenesis. In: Trayhurn P, Nicholls DG (eds) Brown adipose tissue. Edward Arnold, London, pp 31–51

135. Freake HC, Oppenheimer JH (1995) Thermogenesis and thyroid function. Annu Rev Nutr 15: 263–291

136. Fuller A, Carter RN, Mitchell D (1998) Brain and abdominal temperatures at fatigue in rats exercising in the heat. J Appl Physiol 84: 877–883

137. Fuller CA, Baker MA (1983) Selective regulation of brain and body temperatures in the squirrel monkey. Am J Physiol 245: R293–R297

138. Gagge AP, Gonzalez RR (1996) Mechanisms of heat exchange: biophysics and physiology. In: Fregly MJ, Blatteis CM (eds) Handbook of physiology, sect 4: Environmental physiology, vol I. Oxford University Press, New York, pp 45–84

139. Gagge AP, Nishi Y (1977) Heat exchange between human skin surface and thermal environment. In: Lee DHK, Falk HL, Murphy SD, Geiger SR (eds) Handbook of physiology, sect 9: Reactions to environmental agents. American Physiological Society, Bethesda, Md., pp 69–92

140. Gathiram P, Wells MT, Raidoo D, Brock–Utne JG, Gaffin SL (1988) Portal and systemic plasma lipopolysaccharide concentrations in heat–stressed primates. Circ Shock 25: 223–230

141. Gebremedhin KG (1985) Heat exchange between livestock and the environment. In: Yousef MK (ed) Stress physiology in livestock. vol I: Basic principles. CRC Press, Boca Raton, pp 15–33

142. Geiser F (1988) Reduction of metabolism during hibernation and daily torpor in mammals and birds: temperature effect or physiological inhibition? J Comp Physiol (B) 158: 25–37

143. Geiser F, Ruf T (1995) Hibernation versus daily torpor in mammals and birds: physiological variables and classification of torpor patterns. Physiol Zool 68: 935–966

144. Genoud D (1993) Temperature regulation in subtropical tree bats. Comp Biochem Physiol (A) 104: 321–331
145. Geor RJ, McCutcheon LJ (1998) Hydration effects on physiological strain of horses during exercise–heat stress. J Appl Physiol 84: 2042–2051
146. Giesbrecht GG, Arnett JL, Vela E, Bristow GK (1993) Effect of task complexity on mental performance during immersion hypothermia. Aviat Space Environ Med 64: 206–211
147. Gilbert TM, Blatteis CM (1977) Hypothalamic thermoregulatory pathways in the rat. J Appl Physiol 43: 770–777
148. Glotzbach SF, Heller HC (1976) Central nervous regulation of body temperature during sleep. Science 194: 537–539
149. Goldberg MB, Langman VA, Taylor CR (1981) Panting in dogs: paths of air flow in response to heat and exercise. Respir Physiol 43: 327–338
150. Gonzalez RR, Kluger MJ, Hardy JD (1971) Partitional calorimetry of the New Zealand white rabbit at temperatures 5–35°C. J Appl Physiol 31: 728–734
151. González–Alonso J, Calbet JAL, Nielsen B (1998) Muscle blood flow is reduced with dehydration during prolonged exercise in humans. J Physiol (Lond) 513: 895–905
152. González–Alonso J, Teller T, Andersen SL, Jensen FB, Hyldig T, Nielsen B (1999) Influence of body temperature on the development of fatigue during prolonged exercise in the heat. J Appl Physiol 86: 1032–1039
153. Gordon CJ (1990) Thermal biology of the laboratory rat. Physiol Behav 47: 963–991
154. Graham N,McC., Wainman FW, Blaxter KL, Armstrong DG (1958) Environmental temperature, energy metabolism and heat regulation in closely clipped sheep. I. Energy metabolism in closely clipped sheep. J Agric Sci (Camb) 52: 13–24
155. Grant TR (1983) Body temperatures of free–ranging platypuses, Ornithorhynchus anatinus (Monotremata), with observations on their use of burrows. Aust J Zool 31: 117–122
156. Grant TR, Dawson TJ (1978) Temperature regulation in the platypus, Orithorhynchus anatinus: production and loss of metabolic heat in air and water. Physiol Zool 51: 315–332
157. Griffin JD, Kaple ML, Chow AR, Boulant JA (1996) Cellular mechanisms for neuronal thermosensitivity in the rat hypothalamus. J Physiol (Lond) 492: 231–242
158. Grosse M, Jänig W (1976) Vasoconstrictor and pilomotor fibres in skin nerves to the cat's tail. Pflügers Arch 361: 221–229
159. Gundlach H (1967) Brutfürsorge, Brutpflege, Verhaltensontogenese und Tagesperiodik beim Europäischen Wildschwein (Sus scrofa). Z Tierpsychol 25: 955–995
160. Gupta BN, Nier K, Hensel H (1979) Cold–sensitive afferents from the abdomen. Pflügers Arch 380: 203–204
161. Hainsworth FR (1967) Saliva spreading, activity, and body temperature regulation in the rat. Am J Physiol 212: 1288–1292
162. Hales JRS (1973) Effects of exposure to hot environments on the regional distribution of blood flow and on cardiorespiratory function in sheep. Pflügers Arch 344: 133–148
163. Hales JRS (1981) Peripheral effector mechanisms of thermoregulation – regulation of panting. In: Szelenyi Z, Szekely M (eds) Contributions to thermal physiology. Akademiai Kiado, Budapest, pp 421–426
164. Hales JRS (1985) Skin arteriovenous anastomoses, their control and role in thermo-regulation. In: Johansen K, Burggren W (eds) Cardiovascular shunts. Munksgaard, Copenhagen, pp 433–448
165. Hales JRS (1987) Proposed mechanisms underlying heat stroke. In: Hales JRS, Richards DAB (eds) Heat stress: Physical exertion and environment. Elsevier Science Publishers B.V., Amsterdam, pp 85–102
166. Hales JRS, Bennett JW, Fawcett AA (1976) Effects of acute cold exposure on the distri-bution of cardiac output in the sheep. Pflügers Arch 366: 153–157
167. Hales JRS, Findlay JD (1968) Respiration of the ox: normal values and the effects of exposure to hot environments. Respir Physiol 4: 333–352
168. Hales JRS, Findlay JD (1968) The oxygen cost of thermally–induced and CO2–induced hyperventilation in the ox. Respir Physiol 4: 353–362
169. Hales JRS, Hubbard RW, Gaffin SL (1996) Limitation of heat tolerance. In: Fregly MJ, Blatteis CM (eds) Handbook of physiology, sect 4: Environmental physiology, vol I. Oxford University Press, New York, pp 285–355

170. Hales JRS, Iriki M, Tsuchiya K, Kozawa E (1978) Thermally–induced cutaneous sympathetic activity related to blood flow through capillaries and arteriovenous anastomoses. Pflügers Arch 375: 17–24
171. Hales JRS, Jessen C, Fawcett AA, King RB (1985) Skin AVA and capillary dilation and constriction induced by local skin heating. Pflügers Arch 404: 203–207
172. Hales JRS, Khogali M, Fawcett AA, Mustafa MKY (1987) Circulatory changes associated with heat stroke: observations in an experimental animal model. Clin Exp Pharmacol Physiol 14: 761–777
173. Hales JRS, Rowell LB, King RB (1979) Regional distribution of blood flow in awake heat–stressed baboons. Am J Physiol 237: H705–H712
174. Hales JRS, Sakurada S (1998) Heat tolerance. A role for fever? In: Kluger MJ, Bartfei T, Dinarello C (eds) Molecular mechanisms of fever (Annals of the New York Academy of Sciences, vol 856). New York Academy of Sciences, New York, pp 188–205
175. Hallwachs O (1960) Sauerstoffverbrauch und Temperaturverhalten des unnarkotisierten Hundes bei Lufttemperaturen von −10 bis +35°C. Pflügers Arch 271: 748–760
176. Hammel HT (1955) Thermal properties of fur. Am J Physiol 182: 369–376
177. Hammel HT (1964) Terrestrial animals in cold: recent studies of primitive man. In: Dill DB, Adolph EF, Wilber CG (eds) Handbook of physiology, sect 4: Adaptation to the environment. American Physiological Society, Washington, D.C., pp 413–434
178. Hammel HT (1965) Neurons and temperature regulation. In: Yamamoto WS, Brobeck JR (eds) Physiological controls and regulation. Saunders, Philadelphia, pp 71–97
179. Hammel HT (1983) Phylogeny of regulatory mechanisms in temperature regulation. J Therm Biol 8: 37–42
180. Hammel HT, Elsner RW, Heller HC, Maggert J, Bainton CR (1977) Thermoregulatory responses to altering hypothalamic temperature in the harbor seal. Am J Physiol 232: R18–R26
181. Hammel HT, Elsner RW, Messurier DH,Le, Andersen HT, Milan FA (1959) Thermal and metabolic responses of the Australian aborigine exposed to moderate cold in summer. J Appl Physiol 14: 605–615
182. Hammel HT, Wyndham CH, Hardy JD (1958) Heat production and heat loss in the dog at 8–36°C environmental temperature. Am J Physiol 194: 99–108
183. Hardy JD (1961) Physiology of temperature regulation. Physiol Rev 41: 521–606
184. Hardy JD (1972) Models of temperature regulation. In: Bligh J, Moore RE (eds) Essays on temperature regulation. North–Holland, Amsterdam, pp 163–186
185. Hardy JD (1973) Posterior hypothalamus and the regulation of body temperature. Fed Proc 32: 1564–1571
186. Hardy JD, Milhorat AT, Dubois EF (1941) Basal metabolism and heat loss of young women at temperatures from 22°C to 35°C. J Nutr 21: 383–404
187. Hashimoto M, Arita J, Shibata M (1998) Electrical stimulation of the lower midbrain around retrorubral field decreases temperatures of brown fat and rectum in anesthetized Wistar rats. Neurosci Lett 246: 129–132
188. Hayward JN, Baker MA (1969) A comparative study of the role of the cerebral arterial blood in the regulation of brain temperature in five mammals. Brain Res 116: 417–440
189. Hayward JS, Eckerson JD, Collis ML (1977) Thermoregulatory heat production in man: prediction equation based on skin and core temperatures. J Appl Physiol 42: 377–384
190. Heath ME (1985) Effect of cutaneous denervation of face and trunk on thermoregulatory responses to cold in rats. J Appl Physiol 58: 376–383
191. Heath ME, Jessen C (1988) Thermosensitivity of the goat's brain. J Physiol (Lond) 400: 61–74
192. Heldmaier G (1971) Nonshivering thermogenesis and body size in mammals. Z Vergl Physiol 73: 222–248
193. Heldmaier G (1989) Seasonal acclimatization of energy requirements in mammals: functional significance of body weight control, hypothermia, torpor and hibernation. In: Wieser W, Gnaiger E (eds) Energy transformation in cells and organisms. Thieme, Stuttgart, pp 130–139
194. Heldmaier G, Ruf T (1992) Body temperature and metabolic rate during natural hypothermia in endotherms. J Comp Physiol (B) 162: 696–706

195. Heldmaier G, Steinlechner S (1981) Seasonal control of energy requirements for thermo-regulation in the Djungarian hamster (*Phodopus sungorus*), living in natural photoperiod. J Comp Physiol (B) 142: 429–437
196. Heldmaier G, Steinlechner S, Rafael J, Latteier B (1982) Photoperiod and ambient temperature as environmental cues for seasonal thermogenic adaptation in the Djungarian Hamster, *Phodopus sungorus*. Int J Biometeorol 26: 339–345
197. Heldmaier G, Steinlechner S, Rafael J, Vsiansky P (1981) Photoperiodic control and effects of melatonin on nonshivering thermogenesis and brown adipose tissue. Science 212: 917–919
198. Heller HC, Colliver GW, Beard J (1977) Thermoregulation during entrance into hibernation. Pflügers Arch 369: 55–60
199. Heller HC, Edgar DM, Grahn DA, Glotzbach SF (1996) Sleep, thermoregulation, and circadian rhythms. In: Fregly MJ, Blatteis CM (eds) Handbook of physiology, sect 4: Environmental physiology, vol II. Oxford University Press, New York, pp 1361–1374
200. Heller HC, Henderson JA (1976) Hypothalamic thermosensitivity and regulation of storage behavior in a day–active desert rodent *Ammospermophilus nelsoni*. J Comp Physiol (B) 108: 255–270
201. Hellon RF (1975) Monoamines, pyrogens and cations: their actions on central control of body temperature. Pharmacol Rev 26: 289–321
202. Hellon RF (1983) Thermoreceptors. In: Shepherd JT, Abboud FM, Geiger SR (eds) Handbook of physiology, sect 2, vol III/2: Peripheral circulation and organ blood flow. American Physiological Society, Bethesda, Md, pp 659–673
203. Hellon RF, Hensel H, Schäfer K (1975) Thermal receptors in the scrotum of the rat. J Physiol (Lond) 248: 349–357
204. Hellon RF, Misra NK (1973) Neurones in the ventrobasal complex of the rat thalamus responding to scrotal skin temperature changes. J Physiol (Lond) 232: 389–400
205. Hellon RF, Mitchell D (1975) Convergence in a thermal afferent pathway in the rat. J Physiol (Lond) 248: 359–376
206. Hellon RF, Taylor DCM (1982) An analysis of a thermal afferent pathway in the rat. J Physiol (Lond) 326: 319–328
207. Hellon RF, Townsend Y, Laburn HP, Mitchell D (1991) Mechanisms of fever. In: Schönbaum E, Lomax P (eds) Thermoregulation: pathology, pharmacology, and therapy. Pergamon Press, New York, pp 19–54
208. Hellstroem B, Hammel HT (1967) Some characteristics of temperature regulation in the unanesthetized dog. Am J Physiol 213: 547–556
209. Hemingway A (1963) Shivering. Physiol Rev 43: 397–422
210. Hensel H (1973) Neural processes in thermoregulation. Physiol Rev 53: 948–1017
211. Hensel H (1981) Thermoreception and temperature regulation. Academic Press, London, pp 1–321
212. Hensel H (1982) Thermal sensations and thermoreceptors in man. Thomas, Springfield, pp 1–187
213. Hensel H, Andres KH, Düring MV,von (1974) Structure and function of cold receptors. Pflügers Arch 352: 1–11
214. Hensel H, Brück K, Raths P (1973) Homeothermic organisms. In: Precht H, Christopher-sen J, Hensel H, Larcher W (eds) Temperature and life. Springer, Berlin, pp 503–761
215. Hensel H, Kenshalo DR (1969) Warm receptors in the nasal region of cats. J Physiol (Lond) 204: 99–112
216. Henshaw RE, Underwood LS, Casey TM (1972) Peripheral thermoregulation: foot temperature in two arctic canines. Science 175: 988–990
217. Hessemer V, Brück K (1985) Influence of menstrual cycle on thermoregulatory, metabolic, and heart rate responses to exercise at night. J Appl Physiol 59: 1911–1917
218. Hessemer V, Brück K (1985) Influence of menstrual cycle on shivering, skin blood flow, and sweating responses measured at night. J Appl Physiol 59: 1902–1910
219. Himms–Hagen J (1986) Brown adipose tissue and cold–acclimation. In: Trayhurn P, Nicholls DG (eds) Brown adipose tissue. Edward Arnold, London, pp 214–268

220. Himms–Hagen J (1995) Role of brown adipose tissue thermogenesis in control of thermo-regulatory feeding in rats: a new hypothesis that links thermostatic and glucostatic hypotheses for control of food intake. Proc Soc Exp Biol Med 208: 159–169

221. Himms–Hagen J (1996) Neural and hormonal responses to prolonged cold exposure. In: Fregly MJ, Blatteis CM (eds) Handbook of physiology, sect 4: Environmental physiology, vol I. Oxford University Press, New York, pp 439–480

222. Hirata K, Nagasaka T, Noda Y (1989) Venous return from distal regions affects heat loss from the arms and legs during exercise–induced thermal loads. Eur J Appl Physiol 58: 865–872

223. Hislop MS, Buffenstein R (1994) Noradrenaline induces nonshivering thermogenesis in both the naked mole–rat (*Heterocephalus glaber*) and the Damara mole–rat (*Cryptomis damarensis*) despite very different modes of thermoregulation. J Therm Biol 19: 25–32

224. Hissa R (1968) Postnatal development of thermoregulation in the Norwegian Lemming and the Golden Hamster. Ann Zool Fenn 5: 345–383

225. Hochachka PW, Guppy M (1987) Metabolic arrest and the control of biological time. Harvard University Press, Cambridge, Massachusetts, and London, England, pp 1–227

226. Hodgson DR, McCutcheon LJ, Byrd SK, Brown WS, Bayly WM, Brengelmann GL, Gollnick PD (1993) Dissipation of metabolic heat in the horse during exercise. J Appl Physiol 74: 1161–1170

227. Hodgson DR, Rose RJ, Kelso TB, McCutcheon LJ, Bayly WM, Gollnick PD (1990) Respiratory and metabolic response in the horse during moderate and heavy exercise. Pflügers Arch 417: 73–78

228. Hofman WF, Riegle GD (1977) Respiratory evaporative heat loss regulation in shorn and unshorn sheep during mild heat stress. Respir Physiol 30: 339–348

229. Hofmeyr MD, Louw GN (1987) Thermoregulation, pelage conductance and renal function in the desert–adapted springbok, *Antidorcas marsupialis*. Journal of Arid Environments 13: 137–151

230. Hong SK (1973) Pattern of cold adaptation in women divers of Korea (ama). Fed Proc 32: 1614–1622

231. Hong SK, Bennett PB, Shiraki K, Lin YC, Claybaugh JR (1996) Mixed–gas saturation diving. In: Fregly MJ, Blatteis CM (eds) Handbook of physiology, sect 4: Environmental physiology, vol II. Oxford University Press, New York, pp 1023–1045

232. Hopkins P, Knights G (1984) Sweating can be a significant route of evaporative heat loss in sheep. In: Hales JRS (ed) Thermal physiology. Raven, New York, pp 275–278

233. Hori T (1991) An update on thermosensitive neurons in the brain: from cellular biology to thermal and non–thermal homeostatic functions. Jpn J Physiol 41: 1–22

234. Hori T, Harada Y (1976) Midbrain neuronal responses to local and spinal cord temperatures. Am J Physiol 231: 1573–1578

235. Hori T, Kiyohara T, Shibata M, Nakashima T (1984) Involvement of prefrontal cortex in the central control of thermoregulation. In: Hales JRS (ed) Thermal physiology. Raven Press, New York, pp 71–74

236. Horowitz M, Meiri U (1985) Thermoregulatory activity in the rat: effects of hypo-hydration, hypovolemia and hypertonicity and their interaction with short–term heat acclimation. Comp Biochem Physiol (A) 82: 577–582

237. Horwitz BA (1996) Homeostatic responses to acute cold exposure: thermogenic responses in birds and mammals. In: Fregly MJ, Blatteis CM (eds) Handbook of physiology, sect 4: Environmental physiology, vol I. Oxford University Press, New York, pp 359–377

238. Höfler W (1968) Changes in regional distribution of sweating during acclimatization to heat. J Appl Physiol 25: 503–506

239. Hubbard RW (1990) Heat stroke pathophysiology: the energy depletion model. Med Sci Sports Exerc 22: 19–28

240. Hubbard RW, Bowers WD, Matthew WT, Curtis FC, Criss REL, Sheldon GM, Ratteree JW (1977) Rat model of acute heatstroke mortality. J Appl Physiol 42: 809–816

241. Hudson JW (1978) Shallow daily torpor: a thermoregulatory adaptation. In: Wang LCH, Hudson JW (eds) Strategies in cold: natural torpidity and thermogenesis. Academic Press, New York, pp 67–108

242. Hulbert AJ, Dawson TJ (1974) Thermoregulation in perameloid marsupials from different environments. Comp Biochem Physiol (A) 47: 591–616
243. Iggo A (1969) Cutaneous thermoreceptors in primates and sub–primates. J Physiol (Lond) 200: 403–430
244. Ingram DL (1965) Evaporative cooling in the pig. Nature 207: 415–416
245. Ingram DL, Mount LE (1975) Man and animals in hot environments. Springer, Berlin, Heidelberg, New York, pp 1–185
246. Inomoto T, Mercer JB, Simon E (1983) Interaction between hypothalamic and extra-hypothalamic body temperatures in the control of panting in rabbits. Pflügers Arch 398: 142–146
247. Inoue S, Murakami N (1976) Unit responses in the medulla oblongata of rabbit to changes in local and cutaneous temperature. J Physiol (Lond) 259: 339–356
248. Irving L (1956) Physiological insulation of swine as bare–skinned mammals. J Appl Physiol 9: 414–420
249. Irving L (1964) Terrestrial animals in cold: birds and mammals. In: Dill DB, Adolph EF, Wilber CG (eds) Handbook of physiology, sect 4: Adaptation to the environment. American Physiological Society, Washington,D.C., pp 361–377
250. Irving L, Hart JS (1957) The metabolism and insulation of seals as bare–skinned mammals in cold water. Can J Zool 35: 497–511
251. Irving L, Krog J (1955) Temperature of skin in the Arctic as a regulator of heat. J Appl Physiol 7: 355–364
252. Irving L, Peyton LJ, Bahn CH, Peterson RS (1962) Regulation of temperature in fur seals. Physiol Zool 35: 275–284
253. Ivanov KP, Konstantinov V, Danilova N (1982) Thermoreceptor localization in the deep and surface skin layers. J Therm Biol 7: 75–82
254. Jansky L (1995) Humoral thermogenesis and its role in maintaining energy balance. Physiol Rev 75: 237–259
255. Jansky L, Janakova H, Ulicny B, Sramek P, Hosek V, Heller J, Parizkova J (1996) Changes in thermal homeostasis in humans due to repeated cold water immersions. Pflügers Arch 432: 368–372
256. Jenkinson DMcEwan, Robertshaw D (1971) Studies on the nature of sweat gland fatigue in the goat. J Physiol (Lond) 212: 455–465
257. Jessen C (1987) Hyperthermia and its effect on exercise performance. In: Hales JRS, Richards DAB (eds) Physical exertion and environment. Elsevier Science Publishers, Amsterdam, pp 241–249
258. Jessen C (1990) On the location and nature of central thermosensitive structures in homeo-therms. In: Nazar K, Terjung RL, Kaciuba–Uscilko H, Budohoski L (eds) International perspectives in exercise physiology. Human Kinetics Books, Champain, Il., pp 182–187
259. Jessen C (1996) Interaction of body temperatures in control of thermoregulatory effector mechanisms. In: Fregly MJ, Blatteis CM (eds) Handbook of physiology, sect 4: Environmental physiology, vol I. Oxford University Press, New York, pp 127–138
260. Jessen C (1998) Brain cooling: an economy mode of temperature regulation in artio-dactyls. News Physiol Sci 13: 281–286
261. Jessen C, Dmi'el R, Choshniak I, Ezra D, Kuhnen G (1998) Effects of dehydration and rehydration on body temperatures in the black Bedouin goat. Pflügers Arch 436: 659–666
262. Jessen C, Feistkorn G (1983) Some aspects of cutaneous blood flow and acid–base balance during hyperthermia. In: Khogali M, Hales JRS (eds) Heat stroke and temperature regulation. Academic Press, Sydney, pp 241–252
263. Jessen C, Feistkorn G (1984) Some characteristics of core temperature signals in the con-scious goat. Am J Physiol 247: R456–R464
264. Jessen C, Feistkorn G, Nagel A (1983) Temperature sensitivity of skeletal muscle in the conscious goat. J Appl Physiol 54: 880–886
265. Jessen C, Felde D, Volk P, Kuhnen G (1990) Effects of spinal cord temperature on gener-ation and transmission of temperature signals in the goat. Pflügers Arch 416: 428–433
266. Jessen C, Kuhnen G (1992) No evidence for brainstem cooling during face fanning in humans. J Appl Physiol 72: 664–669

267. Jessen C, Laburn HP, Knight MH, Kuhnen G, Goelst K, Mitchell D (1994) Blood and brain temperatures of free–ranging black wildebeest in their natural environment. Am J Physiol 267: R1528–R1536

268. Jessen C, McLean JA, Calvert DT, Findlay JD (1972) Balanced and unbalanced temperature signals generated in spinal cord of the ox. Am J Physiol 222: 1343–1347

269. Jessen C, Mercer JB (1978) Influence of core temperature on respiratory evaporative heat loss in exercising goats. J Physiol (Lond) 284: P162–P163

270. Johansson BW (1996) The hibernator heart – Nature's model of resistance to ventricular fibrillation. Cardiovasc Res 31: 826–832

271. Johnsen HK, Blix AS, Jorgensen L, Mercer JB (1985) Vascular basis for regulation of nasal heat exchange in reindeer. Am J Physiol 249: R617–R623

272. Johnsen HK, Folkow LP (1988) Vascular control of brain cooling in reindeer. Am J Physiol 254: R730–R739

273. Johnsen HK, Rognmo A, Nilssen KJ, Blix AS (1985) Seasonal changes in the relative importance of different avenues of heat loss in resting and running reindeer. Acta Physiol Scand 123: 73–79

274. Johnson JM, Proppe DW (1996) Cardiovascular adjustments to heat stress. In: Fregly MJ, Blatteis CM (eds) Handbook of physiology, sect 4: Environmental physiology, vol I. Oxford University Press, New York, pp 215–243

275. Johnson JM, Rowell LB, Brengelmann GL (1974) Modification of the skin blood flow–body temperature relationship by upright exercise. J Appl Physiol 37: 880–886

276. Johnson KG, Strack R (1989) Adaptive behaviour of laboratory rats feeding in hot conditions. Comp Biochem Physiol (A) 94: 69–72

277. Kanosue K, Nakayama T, Tanaka H, Yanase M, Yasuda H (1990) Modes of action of local hypothalamic and skin thermal stimulation on salivary secretion in rats. J Physiol (Lond) 424: 459–472

278. Kanosue K, Niwa K, Andrew PD, Yasuda H, Yanase M, Tanaka H, Matsumura K (1991) Lateral distribution of hypothalamic signals controlling thermoregulatory vasomotor activity and shivering in rats. Am J Physiol 260: R486–R493

279. Kanosue K, Yanase–Fujiwara M, Hosono T (1994) Hypothalamic network for thermoregulatory vasomotor control. Am J Physiol 267: R283–R288

280. Kanosue K, Zhang Y–H, Yanase–Fujiwara M, Hosono T (1994) Hypothalamic network for thermoregulatory shivering. Am J Physiol 267: R275–R282

281. Keatinge WR, Sloan REG (1975) Deep body temperature from aural canal with servo–controlled heating to outer ear. J Appl Physiol 38: 919–921

282. Keller AD, McClaskey EB (1964) Localization, by the brain slicing method, of the level or levels of the cephalic brainstem upon which effective heat dissipation is dependent. Am J Phys Med 43: 181–213

283. Kellogg DL,Jr., Crandall CG, Liu Y, Charkoudian N, Johnson JM (1998) Nitric oxide and cutaneous active vasodilation during heat stress in humans. J Appl Physiol 85: 824–829

284. Kellogg DL,Jr., Liu Y, Kosiba IF, O'Donnell D (1999) Role of nitric oxide in the vascular effects of local warming of the skin in humans. J Appl Physiol 86: 1185–1190

285. Kenshalo R (1990) Correlations of temperature sensation and neural activity: a second approximation. In: Bligh J, Voigt K (eds) Thermoreception and temperature regulation. Springer, Berlin, pp 215–243

286. Kerr MG, Snow DH (1983) Composition of sweat of the horse during prolonged epinephrine (adrenaline) infusion, heat exposure, and exercise. Am J Vet Res 44: 1571–1577

287. Kerslake DMcK (1972) The stress of hot environments. University Press, Cambridge, pp 1–316

288. Khogali M (1983) Heat stroke: an overview. In: Khogali M, Hales JRS (eds) Heat stroke and temperature regulation. Academic Press, Sydney, pp 1–12

289. Kleiber M (1961) The Fire of Life: an introduction to animal energetics. Wiley, New York, pp 1–454

290. Kleinebeckel D, Klußmann FW (1990) Shivering. In: Schönbaum E, Lomax P (eds) Thermoregulation: physiology and pharmacology. Pergamon Press, New York, pp 235–254

291. Kluger MJ (1975) Fever. Princeton University Press, Princeton, NJ, pp 1–195

292. Kondo N, Tominaga H, Shibasaki M, Aoki K, Koga S, Nishiyasu T (1999) Modulation of the thermoregulatory sweating response to mild hyperthermia during activation of the muscle metaboreflex in humans. J Physiol (Lond) 515: 591–598

293. Kozak W, Kluger MJ, Soszynski D, Conn CA, Rudolph K, Leon LR, Zheng H (1998) IL–6 and IL–1ß in fever. Studies using cytokine–deficient (knockout) mice. In: Kluger MJ, Bartfei T, Dinarello C (eds) Molecular mechanisms of fever (Annals of the New York Academy of Sciences, vol 856). New York Academy of Sciences, New York, pp 33–47

294. Krönert H, Wurster RD, Pierau FK, Pleschka K (1980) Vasodilatory response of arterio-venous anastomoses to local cold stimuli in the dog's tongue. Pflügers Arch 388: 17–19

295. Kuhnen G (1997) Selective brain cooling reduces respiratory water loss during heat stress. Comp Biochem Physiol (A) 118: 891–895

296. Kuhnen G, Jessen C (1988) The metabolic response to skin temperature. Pflügers Arch 412: 402–408

297. Kuhnen G, Jessen C (1991) Threshold and slope of selective brain cooling. Pflügers Arch 418: 176–183

298. Kuhnen G, Jessen C (1994) Thermal signals in control of selective brain cooling. Am J Physiol 267: R355–R359

299. Laburn HP, Mitchell D, Goelst K (1992) Fetal and maternal body temperatures measured by radiotelemetry in near–term sheep during thermal stress. J Appl Physiol 72: 894–900

300. Lange Andersen K, Loyning Y, Nelms JD, Wilson O, Fox RH, Bolstad A (1960) Metabolic and thermal response to a moderate cold exposure in nomadic Lapps. J Appl Physiol 15: 649–653

301. Langman VA, Maloiy GMO (1989) Passive obligatory heterothermy of the giraffe. J Physiol (Lond) 415: 89P (Abstract)

302. Langman VA, Maloiy GMO, Schmidt–Nielsen K, Schroter RC (1979) Nasal heat ex-changer in the giraffe and other large mammals. Respir Physiol 37: 325–333

303. Leblanc J, Hildes JA, Heroux O (1960) Tolerance of Gaspe fishermen to cold water. J Appl Physiol 15: 1031–1034

304. Lentz CP, Hart JS (1960) The effect of wind and moisture on heat loss through the fur of newborn caribou. Can J Zool 38: 679–688

305. Li H, Satinoff E (1998) Fetal tissue containing the suprachiasmatic nucleus restores mul-tiple circadian rhythms in old rats. Am J Physiol 275: R1735–R1744

306. Lin MT, Simon E (1982) Properties of high Q10 units in the conscious duck's hypo-thalamus responsive to changes of core temperature. J Physiol (Lond) 322: 127–137

307. Lindstroem S, Mazières L (1991) Effect of menthol on the bladder cooling reflex in the cat. Acta Physiol Scand 141: 1–10

308. Lipton JM (1968) Effects of preoptic lesions on heat–escape responding and colonic tem-perature in the rat. Physiol Behav 3: 165–169

309. Lipton JM (1971) Thermal stimulation of the medulla alters behavioral temperature regu-lation. Brain Res 26: 439–442

310. Lipton JM (1973) Thermosensitivity of medulla oblongata in control of body temperature. Am J Physiol 224: 890–897

311. Louw G, Seely M (1982) Ecology of desert organisms. Longman, London, pp 1–194

312. Lovegrove BG, Heldmaier G, Ruf T (1991) Perspectives of endothermy revisited: the endothermic temperature range. J Therm Biol 16: 185–197

313. Lund RJ, Guthrie AJ, Moster HJ, Travers CW, Nurton JP, Adamson DJ (1996) Effect of three different warm–up regimens on heat balance and oxygen consumption of thorough-bred horses. J Appl Physiol 80: 2190–2197

314. MacArthur RA (1989) Aquatic mammals in cold. In: Wang LCH (ed) Advances in com-parative and environmental physiology, vol 4. Springer, Berlin Heidelberg, pp 289–325

315. Macfarlane WV (1964) Terrestrial animals in dry heat: ungulates. In: Dill DB, Adolph EF, Wilber CG (eds) Handbook of physiology, sect 4: Adaptation to the environment. American Physiological Society, Washington,D.C., pp 509–539

316. Mack GW, Nadel ER (1996) Body fluid balance during heat stress in humans. In: Fregly MJ, Blatteis CM (eds) Handbook of physiology, sect 4: Environmental physiology, vol I. Oxford University Press, New York, pp 187–214

317. Mack GW, Nose H, Nadel ER (1988) Role of cardiopulmonary baroreflexes during dynamic exercise. J Appl Physiol 65: 1827–1832
318. MacMillin JM, Seal US, Karns PD (1980) Hormonal correlates of hypophagia in white–tailed deer. Fed Proc 39: 2964–2968
319. MacPherson RK (1959) The effect of fever on temperature regulation in man. Clin Sci 18: 281–287
320. Mahoney SA (1980) Cost of locomotion and heat balance during rest and running from 0 to 55°C in a patas monkey. J Appl Physiol 49: 789–800
321. Malan A (1986) pH as a control factor in hibernation. In: Heller HC, Musacchia XJ, Wang LCH (eds) Living in the cold. Elsevier, New York, pp 61–70
322. Maloney SK, Mitchell D (1996) Regulation of ram scrotal temperature during heat exposure, cold exposure, fever and exercise. J Physiol (Lond) 496: 421–430
323. Maron MB, Wagner JA, Horvath SM (1977) Thermoregulatory responses during competitive marathon running. J Appl Physiol 42: 909–914
324. McConaghy FF, Hales JRS, Rose RJ, Hodgson DR (1995) Selective brain cooling in the horse during exercise and environmental heat stress. J Appl Physiol 79: 1849–1854
325. McEwen GN,Jr., Heath JE (1973) Resting metabolism and thermoregulation in the unrestrained rabbit. J Appl Physiol 35: 884–886
326. McLean JA (1974) Loss of heat by evaporation. In: Monteith JL, Mount LE (eds) Heat loss from animals and man. Butterworths, London, pp 19–31
327. McLean JA, Downie AJ, Watts PR, Glasbey CA (1982) Heat balance of ox steers (*Bos taurus*) in steady–temperature environments. J Appl Physiol 52: 324–332
328. McLean JA, Stombaugh DP, Downie AJ, Glasbey CA (1983) Body heat storage in steers (*Bos taurus*) in fluctuating thermal environments. J Agric Sci (Camb) 100: 315–322
329. McLean JA, Tobin G (1987) Animal and human calorimetry. Cambridge University Press, Cambridge, pp 1–338
330. McNab BK (1980) On estimating thermal conductance in endotherms. Physiol Zool 53: 145–156
331. Meeh K (1879) Oberflächenmessungen des menschlichen Körpers. Zeitschrift für Biologie 15: 425–458
332. Mense S, Meyer H (1985) Different types of slowly conducting afferent units in cat skeletal muscle and tendon. J Physiol (Lond) 363: 403–418
333. Mercer JB, Hammel HT (1989) Total calorimetry and temperature regulation in the nine–banded armadillo. Acta Physiol Scand 135: 579–589
334. Mercer JB, Jessen C (1978) Effects of total body core cooling on heat production in conscious goats. Pflügers Arch 373: 259–267
335. Mercer JB, Jessen C (1978) Central thermosensitivity in conscious goats: hypothalamus and spinal cord versus residual inner body. Pflügers Arch 374: 179–186
336. Mitchell D (1974) Convective heat transfer from man and other animals. In: Monteith JL, Mount LE (eds) Heat loss from animals and man. Butterworths, London, pp 59–76
337. Mitchell D, Laburn HP (1997) Macrophysiology of fever. In: Nielsen Johannsen B, Nielsen R (eds) Thermal physiology 1997. The August Krogh Institute, Copenhagen, pp 249–263
338. Mitchell D, Laburn HP, Nijland MJM, Zurovsky Y, Mitchell G (1987) Selective brain cooling and survival. S Afr J Sci 83: 598–604
339. Mitchell D, Maloney SK, Laburn HP, Knight MH, Kuhnen G, Jessen C (1997) Activity, blood temperature and brain temperature of free–ranging springbok. J Comp Physiol (B) 167: 335–343
340. Mitchell D, Senay LC, Wyndham CH, Rensburg AJ,van, Rogers GG, Strydom NB (1976) Acclimatization in a hot, humid environment: energy exchange, body temperature, and sweating. J Appl Physiol 40: 768–778
341. Mitchell D, Snellen JW, Atkins AR (1970) Thermoregulation during fever: change of set–point or change of gain. Pflügers Arch 321: 293–302
342. Mitchell D, Wyndham CH, Atkins AR, Vermeulen AJ, Hofmeyr HS, Strydom NB, Hodgson T (1968) Direct measurement of the thermal responses of nude resting men in dry environments. Pflügers Arch 303: 324–343

343. Mitchell GW (1991) Rapid onset of severe heat illness: a case report. Aviat Space Environ Med 62: 779–782

344. Mittleman KD, Mekjavic IB (1991) Contribution of core cooling rate to shivering thermogenesis during cold water immersion. Aviat Space Environ Med 62: 842–848

345. Moen AN (1978) Seasonal changes in heart rates, activity, metabolism, and forage intake of white–tailed deer. J Wildl Manage 42: 715–738

346. Molinari HH, Kenshalo DR (1977) Effect of cooling rate on the dynamic response of cat cold units. Exp Neurol 55: 546–555

347. Molnar GW (1946) Survival of hypothermia by men immersed in the ocean. J Am Med Assoc 131: 1046–1050

348. Molyneux GS, Bryden MM (1981) Comparative aspects of arteriovenous anastomoses. In: Harrison RJ, Holmes RL (eds) Progress in anatomy. Cambridge University Press, Cambridge, pp 207–227

349. Monteith JL (1975) Principles of environmental physics. Arnold, London, pp 1–241

350. Morimoto T (1990) Thermoregulation and body fluids: role of blood volume and central venous pressure. Jpn J Physiol 40: 165–179

351. Morrison P (1966) Insulative flexibility in the guanaco. J Mammal 47: 18–23

352. Morrison PR, Ryser FA (1952) Weight and body temperature in mammals. Science 116: 231–232

353. Moseley PL (1997) Heat shock proteins and heat adaptation of the whole organism. J Appl Physiol 83: 1413–1417

354. Mount LE (1960) The influence of huddling and body size on the metabolic rate of the young pig. J Agric Sci (Camb) 55: 101–105

355. Mount LE (1974) The concept of thermal neutrality. In: Monteith JL, Mount LE (eds) Heat loss from animals and man. Butterworths, London, pp 425–439

356. Mower GD (1976) Perceived intensity of peripheral thermal stimuli is independent of internal body temperature. J Comp Physiol Psychol 90: 1152–1155

357. Nadel ER (1986) Non–thermal influences on the control of skin blood flow have minimal effects on heat transfer during exercise. Yale J Biol Med 59: 321–327

358. Nadel ER, Bullard RW, Stolwijk JAJ (1971) Importance of skin temperature in the regulation of sweating. J Appl Physiol 31: 80–87

359. Nadel ER, Cafarelli E, Roberts MF, Wenger CB (1979) Circulatory regulation during exercise in different ambient temperatures. J Appl Physiol 46: 430–437

360. Nadel ER, Holmer I, Bergh U, Astrand PO, Stolwijk JAJ (1974) Energy exchanges of swimming man. J Appl Physiol 36: 456–471

361. Nadel ER, Mack GW, Takamata A (1993) Thermoregulation, exercise and thirst: interrelationships in humans. In: Gisolfi CV, Lamb DR, Nadel ER (eds) Exercise, Heat, and Thermoregulation. Brown & Benchmark, Dubuque, IA, pp 225–251

362. Nadel ER, Mitchell GW, Stolwijk JAJ (1973) Differential thermal sensitivity in the human skin. Pflügers Arch 340: 71–76

363. Nadel ER, Stolwijk JAJ (1973) Effects of skin wettedness on sweat gland response. J Appl Physiol 35: 689–694

364. Nagasaka T, Cabanac M, Hirata K, Nunomura T (1986) Heat induced vasoconstriction in the fingers: A mechanism for reducing heat gain through the hand heated locally. Pflügers Arch 407: 71–75

365. Nagel A (1991) Metabolic, respiratory and cardiac activity in the shrew *Crocidura russula*. Respir Physiol 85: 139–149

366. Nagel A, Herold W, Roos U, Jessen C (1986) Skin and core temperatures as determinants of heat production and heat loss in the goat. Pflügers Arch 406: 600–607

367. Nakashima T, Hori T, Kiyohara T, Shibata M (1985) Osmosensitivity of preoptic thermosensitive neurons in hypothalamic slices in vitro. Pflügers Arch 405: 112–117

368. Nakayama T, Hardy JD (1969) Unit responses in the rabbit's brain stem to changes in brain and cutaneous temperature. J Appl Physiol 27: 848–857

369. Necker R (1981) Thermoreception and temperature regulation in homeothermic vertebrates. In: Autrum H, Ottoson D, Perl E, Schmidt RF (eds) Progress in sensory physiology 2. Springer, Heidelberg, pp 1–47

370. Needham AD, Dawson TJ, Hales JRS (1974) Forelimb blood flow and saliva spreading in the thermoregulation of the red kangaroo (*Megaleia rufa*). Comp Biochem Physiol (A) 49: 555–565
371. Nelms JD, Soper DJG (1962) Cold vasodilatation and cold acclimatization in the hands of British fish filleters. J Appl Physiol 17: 444–448
372. Nicholls DG, Cunningham SA, Rial E (1986) The bioenergetic mechanisms of brown adipose tissue mitochondria. In: Trayhurn P, Nicholls DG (eds) Brown adipose tissue. Eward Arnold, London, pp 52–85
373. Nielsen B (1974) Effects of changes in plasma volume and osmolarity on thermoregulation during exercise. Acta Physiol Scand 90: 725–730
374. Nielsen B (1976) Metabolic reactions to changes in core and skin temperature in man. Acta Physiol Scand 97: 128–138
375. Nielsen B (1984) The effect of dehydration on circulation and temperature regulation during exercise. J Therm Biol 9: 107–112
376. Nielsen B, Hales JRS, Strange S, Christensen NJ, Warberg J, Saltin B (1993) Human circulatory and thermoregulatory adaptations with heat acclimation and exercise in a hot, dry environment. J Physiol (Lond) 460: 467–485
377. Nielsen B, Rowell LB, Bonde–Petersen F (1984) Cardiovascular responses to heat stress and blood volume displacements during exercise in man. Eur J Appl Physiol 52: 370–374
378. Nielsen B, Strange S, Christensen NJ, Warberg J, Saltin B (1997) Acute and adaptive responses in humans to exercise in a warm, humid environment. Pflügers Arch 434: 49–56
379. Nielsen M (1970) Heat production and body temperature during rest and work. In: Hardy JD, Gagge AP, Stolwijk JAJ (eds) Physiological and behavioral temperature regulation. Thomas, Springfield, pp 205–214
380. Nilssen KJ, Sundsfjord A, Blix AS (1984) Regulation of metabolic rate in Svalbard and Norwegian reindeer. Am J Physiol 247: R837–R841
381. Nold JL, Peterson LJ, Fedde MR (1991) Physiological changes in the running greyhound (Canis domesticus): influence of race length. Comp Biochem Physiol (A) 100: 623–627
382. Nose H, Mack GW, Shi X, Nadel ER (1988) Shift in body fluid compartments after dehydration in humans. J Appl Physiol 65: 318–324
383. Nose H, Morita M, Yawata T, Morimoto T (1986) Recovery of blood volume and osmolality after thermal dehydration in rats. Am J Physiol 251: R492–R498
384. Ogawa T, Sugenoya J (1993) Pulsatile sweating and sympathetic sudomotor activity. Jpn J Physiol 43: 275–289
385. Ohnishi N, Ogawa T, Sugenoya J, Natsume K, Yamashita Y, Imamura R, Imai K (1994) Central motor commands affects the sweating activity during exercise. In: Milton AS (ed) Temperature regulation: recent physiological and pharmacological advances. Birkhäuser, Basel, pp 183–187
386. Pabst DA, Rommel SA, McLellan WA (1998) Evolution of thermoregulatory function in Cetacean reproductive systems. In: Thewissen JGM (ed) The emergence of whales. Plenum Press, New York, pp 379–397
387. Park YS, Hong SK (1991) Physiology of cold–water diving as exemplified by Korean women divers. Undersea Biomed Res 18: 229–241
388. Parmeggiani PL, Rabini C (1967) Shivering and panting during sleep. Brain Res 6: 789–791
389. Paton BC (1992) Accidental hypothermia. In: Schönbaum E, Lomax P (eds) Thermoregulation: pathology, pharmacology, and therapy. Pergamon Press, New York, pp 397–454
390. Pehl U, Schmid HA, Simon E (1997) Temperature sensitivity of neurones in slices of the rat spinal cord. J Physiol (Lond) 498: 483–495
391. Pérgola PE, Kellogg DL,Jr., Johnson JM, Kosiba WA, Solomon DE (1993) Role of sympathetic nerves in the vascular effects of local temperature in human forearm skin. Am J Physiol 265: H785–H792
392. Pierau FK (1996) Peripheral thermosensors. In: Fregly MJ, Blatteis CM (eds) Handbook of physiology, sect 4: Environmental physiology, vol I. Oxford University Press, New York, pp 85–104

393. Pirlet K (1962) Die Verstellung des Kerntemperatur–Sollwertes bei Kältebelastung. Pflügers Arch 275: 71–94
394. Pleschka K (1984) Control of tongue blood flow in regulation of heat loss in mammals. Rev Physiol Biochem Pharmacol 100: 76–120
395. Pozos RS, Iaizzo PA, Danzl DF, Mills WT,Jr. (1996) Limits of tolerance to hypothermia. In: Fregly MJ, Blatteis CM (eds) Handbook of physiology, sect 4: Environmental physiology, vol I. Oxford University Press, New York, pp 557–575
396. Pugh LGC, Edholm OG (1955) The physiology of Channel swimmers. Lancet 761–768
397. Puschmann S, Jessen C (1978) Anterior and posterior hypothalamus: effects of independent temperature displacements on heat production in conscious goats. Pflügers Arch 373: 59–68
398. Quinton PM (1987) Physiology of sweat secretion. Kidney Int 32, Suppl. 21: S102–S108
399. Raman ER, Roberts MF, Vanhuyse VJ (1983) Body temperature control of rat tail blood flow. Am J Physiol 245: R426–R432
400. Rautenberg W (1981) The importance of pilomotor response in temperature regulation. In: Szelenyi Z, Szekely M (eds) Contributions to thermal physiology. Akademiai Kiado, Budapest, pp 391–395
401. Rawson RO, Quick KP (1972) Localization of intra–abdominal thermoreceptors in the ewe. J Physiol (Lond) 222: 665–677
402. Reeves RB (1977) The interaction of body temperature and acid–base balance in ectothermic vertebrates. Annu Rev Physiol 39: 559–586
403. Refinetti R (1996) Rhythms of body temperature and temperature selection are out of phase in a diurnal rodent, *Octogon degus*. Physiol Behav 60: 959–961
404. Refinetti R, Menaker M (1992) The circadian rhythm of body temperature. Physiol Behav 51: 613–637
405. Refinetti R, Menaker M (1992) Body temperature rhythm of the tree shrew, *Tupaia belangeri*. J Exp Zool 263: 453–457
406. Richards SA (1970) The biology and comparative physiology of thermal panting. Biol Rev Camb Philos Soc 45: 223–264
407. Ricquier D, Bouillaud F (1986) The brown adipose tissue mitochondrial uncoupling protein. In: Trayhurn P, Nicholls DG (eds) Brown adipose tissue. Edward Arnold, London, pp 86–104
408. Riedel W (1976) Warm receptors in the dorsal abdominal wall of the rabbit. Pflügers Arch 361: 205–206
409. Riedel W, Iriki M, Simon E (1972) Regional differentiation of sympathetic activity during peripheral heating and cooling in anesthetized rabbits. Pflügers Arch 332: 239–247
410. Riedel W, Siaplauras G, Simon E (1973) Intra–abdominal thermosensitivity in the rabbit as compared with spinal thermosensitivity. Pflügers Arch 340: 59–70
411. Roberts JC (1996) Thermogenic responses to prolonged cold exposure: birds and mammals. In: Fregly MJ, Blatteis CM (eds) Handbook of physiology, sect 4: Environmental physiology, vol I. Oxford University Press, New York, pp 399–418
412. Roberts MF, Wenger CB, Stolwijk JAJ, Nadel ER (1977) Skin blood flow and sweating changes following exercise training and heat acclimation. J Appl Physiol 43: 133–137
413. Roberts MF, Zygmunt AC (1984) Reflex and local thermal control of rabbit ear blood flow. Am J Physiol 246: R979–R984
414. Roberts WW, Martin JR (1977) Effects of lesions in central thermosensitive areas on thermoregulatory responses in rat. Physiol Behav 19: 503–511
415. Roberts WW, Mooney RD (1974) Brain areas controlling thermoregulatory grooming, prone extension, locomotion, and tail vasodilation in rats. J Comp Physiol Psychol 86: 470–480
416. Robertshaw D (1975) Catecholamines and control of sweat glands. In: Blaschko H, Sayers G, Smith AD (eds) Handbook of physiology, sect 7, vol VI: Adrenal gland. American Physiological Society, Washington, D.C., pp 591–603
417. Robertshaw D, Taylor CR (1969) Sweat gland function of the donkey (*Equus asinus*). J Physiol (Lond) 205: 79–89
418. Robertshaw D, Taylor CR (1969) A comparison of sweat gland activity in eight species of east African bovids. J Physiol (Lond) 203: 135–143

419. Robertshaw D, Taylor CR, Mazzia LM (1973) Sweating in primates: secretion by adrenal medulla during exercise. Am J Physiol 224: 678–681
420. Robinson EL, Fuller CA (1999) Endogenous thermoregulatory rhythms of squirrel monkeys in thermoneutrality and cold. Am J Physiol 276: R1397–R1407
421. Roos U, Jessen C (1986) No dynamic effector responses to fast changes of core temperature at constant skin temperature. Can J Physiol Pharmacol 65: 1339–1346
422. Rowell LB (1974) Human cardiovascular adjustments to exercise and thermal stress. Physiol Rev 54: 75–159
423. Rowell LB (1983) Cardiovascular adjustments to thermal stress. In: Shepherd JT, Abboud FM, Geiger SR (eds) Handbook of physiology, sect 2, vol III/2: Peripheral circulation and organ blood flow. American Physiological Society, Bethesda, Md, pp 967–1023
424. Rowell LB, Brengelmann GL, Blackmon JR, Twiss RD, Kusumi F (1968) Splanchnic blood flow and metabolism in heat–stressed man. J Appl Physiol 24: 475–484
425. Ruf T, Heldmaier G (1992) The impact of daily torpor on energy requirements in the Djungarian hamster, *Phodopus sungorus*. Physiol Zool 65: 994–1010
426. Sakurada S, Hales JRS (1998) A role for gastrointestinal endotoxins in the enhancement of heat tolerance by physical fitness. J Appl Physiol 84: 207–214
427. Satinoff E (1978) Neural organization and evolution of thermal regulation in mammals. Science 201: 16–20
428. Satinoff E (1996) Behavioral thermoregulation in the cold. In: Fregly MJ, Blatteis CM (eds) Handbook of physiology, sect 4: Environmental physiology, vol I. Oxford University Press, New York, pp 481–505
429. Satinoff E, Liran J, Clapman R (1982) Aberrations of circadian body temperature rhythms in rats with medial preoptic lesions. Am J Physiol 242: R352–R357
430. Satinoff E, Rutstein J (1970) Behavioral thermoregulation in rats with anterior hypothalamic lesions. J Comp Physiol Psychol 71: 77–82
431. Satinoff E, Shan S (1971) Loss of behavioral thermoregulation after lateral hypothalamic lesions in rats. J Comp Physiol Psychol 77: 302–312
432. Sato F, Owen M, Matthes R, Sato K, Gisolfi CV (1990) Functional and morphological changes in the eccrine sweat gland with heat acclimation. J Appl Physiol 69: 232–236
433. Sato K (1993) The mechanism of eccrine sweat secretion. In: Gisolfi CV, Lamb DR, Nadel ER (eds) Perspectives in exercise science and sports medicine, vol 6: Exercise, heat, and thermoregulation. Brown & Benchmark, Dubuque, IA, pp 85–110
434. Sawka MN, Wenger CB, Pandolf KB (1996) Thermoregulatory responses to acute exercise–heat stress and heat acclimation. In: Fregly MJ, Blatteis CM (eds) Handbook of physiology, sect 4: Environmental physiology, vol I. Oxford University Press, New York, pp 157–185
435. Sawka MN, Young AJ, Francesconi RP, Muza SR, Pandolf KB (1985) Thermoregulatory and blood responses during exercise at graded hypohydration levels. J Appl Physiol 59: 1394–1401
436. Scales WE, Kluger MJ (1987) Effect of antipyretic drugs on circadian rhythm in body temperature of rats. J Appl Physiol 253: R306–R313
437. Schmidt I (1978) Interactions of behavioural and autonomic thermoregulation in heat stressed pigeons. Pflügers Arch 374: 47–55
438. Schmidt I (1984) Interaction of behavioral and autonomic thermoregulation. In: Hales JRS (ed) Thermal physiology. Raven Press, New York, pp 309–318
439. Schmidt–Nielsen K (1964) Terrestrial animals in dry heat: desert rodents. In: Dill DB, Adolph EF, Wilber CG (eds) Handbook of physiology, sect 4: Adaptation to the environment. American Physiological Society, Washington,D.C., pp 493–507
440. Schmidt–Nielsen K, Crawford EC,Jr., Newsome AE, Rawson KS, Hammel HT (1967) Metabolic rate of camels: effect of body temperature and dehydration. Am J Physiol 212: 341–346
441. Schmidt–Nielsen K, Dawson TJ, Hammel HT, Hinds D, Jackson DC (1965) The Jack rabbit – a study in its desert survival. Hvalradets Skrifter 48: 125–142
442. Schmidt–Nielsen K, Hainsworth FR, Murrish DE (1970) Counter–current heat exchange in the respiratory passages: effect on water and heat balance. Respir Physiol 9: 263–276

443. Schmidt–Nielsen K, Schmidt–Nielsen B (1952) Water metabolism of desert mammals. Physiol Rev 32: 135–166
444. Schmidt–Nielsen B, Schmidt–Nielsen K, Houpt TR, Jarnum SA (1956) Water balance of the camel. Am J Physiol 185: 185–194
445. Schmidt–Nielsen K, Schmidt–Nielsen B, Jarnum SA, Houpt TR (1957) Body temperature of the camel and its relation to water economy. Am J Physiol 188: 103–112
446. Schmidt–Nielsen K, Schroter RC, Shkolnik A (1981) Desaturation of exhaled air in camels. Proc R Soc Lond (Biol) 211: 305–319
447. Schmieg G, Mercer JB, Jessen C (1980) Thermosensitivity of the extrahypothalamic brain stem in conscious goats. Brain Res 188: 383–397
448. Scholander PF, Hock R, Walters V, Johnson F, Irving L (1950) Heat regulation in some arctic and tropical mammals and birds. Biol Bull 99: 237–258
449. Scholander PF, Schevill WE (1955) Counter–current vascular heat exchange in the fins of whales. J Appl Physiol 8: 279–282
450. Scholander PF, Walters V, Hock R, Irving L (1950) Body insulation of some arctic and tropical mammals and birds. Biol Bull 99: 225–236
451. Schroter RC, Robertshaw D, Baker MA, Shoemaker VH, Holmes R, Schmidt–Nielsen K (1987) Respiration in heat stressed camels. Respir Physiol 70: 97–112
452. Sehic E, Li S, Ungar AL, Blatteis CM (1998) Complement reduction impairs the febrile response of guinea pigs to endotoxin. Am J Physiol 274: R1594–R1603
453. Setchell BP, Mieusset R (1996) Testis thermoregulation. In: Hamamah S, Mieusset R (eds) Recherches en male gametes: production and quality. Inserm, pp 65–81
454. Shaffrath JD, Adams WC (1984) Effects of airflow and work load on cardiovascular drift and skin blood flow. J Appl Physiol 56: 1411–1417
455. Sharp FR, Hammel HT (1972) Effects of fever on salivation response in the resting and exercising dog. Am J Physiol 223: 77–82
456. Shibolet S, Lancaster MC, Danon Y (1976) Heat stroke: a review. Aviat Space Environ Med 47: 280–301
457. Shido S, Romanovsky AA, Ungar AL, Blatteis CM (1993) Role of intrapreoptic norepinephrine in endotoxin–induced fever in guinea pigs. Am J Physiol 265: R1369–R1375
458. Silva JE, Larsen PR (1985) Potential of brown adipose tissue type II thyroxine 5'–deiodinase as a local and systemic source of triiodothyronine in rats. J clin Invest 76: 2296–2305
459. Silva NL, Boulant JA (1984) Effects of osmotic pressure, glucose, and temperature on neurons in preoptic tissue slices. Am J Physiol 247: R335–R345
460. Simon E (1972) Temperature signals from skin and spinal cord converging on spinothalamic neurons. Pflügers Arch 337: 323–332
461. Simon E (1974) Temperature regulation: the spinal cord as a site of extrahypothalamic thermoregulatory functions. Rev Physiol Biochem Pharmacol 71: 1–76
462. Simon E (1981) Effects of CNS temperature on generation and transmission of temperature signals in homeotherms. A common concept for mammalian and avian thermoregulation. Pflügers Arch 392: 79–88
463. Simon E, Hammel HT, Oksche A (1977) Thermosensitivity of single units in the hypothalamus of the conscious Pekin duck. J Neurobiol 8: 523–535
464. Simon E, Iriki M (1971) Sensory transmission of spinal heat and cold sensitivity in ascending spinal neurons. Pflügers Arch 328: 103–120
465. Simon E, Klußmann FW, Rautenberg W, Kosaka M (1966) Kältezittern bei narkotisierten spinalen Hunden. Pflügers Arch 291: 187–204
466. Simon E, Pierau FK, Taylor DCM (1986) Central and peripheral thermal control of effectors in homeothermic temperature regulation. Physiol Rev 66: 235–300
467. Simon E, Rautenberg W, Thauer R, Iriki M (1964) Die Auslösung von Kältezittern durch lokale Kühlung im Wirbelkanal. Pflügers Arch 281: 309–331
468. Simon E, Riedel W (1975) Diversity of regional sympathetic outflow in integrative cardiovascular control: patterns and mechanisms. Brain Res 87: 323–333
469. Simon–Oppermann C, Simon E, Jessen C, Hammel HT (1978) Hypothalamic thermosensitivity in conscious Pekin ducks. Am J Physiol 235: R130–R140
470. Singer D, Bretschneider HJ (1990) Metabolic reduction in hypothermia: pathophysiological and natural examples – part 2. Thorac Cardiovasc Surgeon 38: 212–219

471. Singer D, Bretschneider HJ (1990) Metabolic reduction in hypothermia: pathophysiological problems and natural examples – part 1. Thorac Cardiovasc Surgeon 38: 205–211
472. Slovis CM, Anderson GF, Casolaro A (1982) Survival in a heat stroke victim with a core temperature in excess of 46.5°C. Ann Emerg Med 11: 269–271
473. Smith JH, Robinson S, Pearcy M (1952) Renal responses to exercise, heat and dehydration. J Appl Physiol 4: 659–665
474. Smith RE, Horwitz BA (1969) Brown fat and thermogenesis. Physiol Rev 49: 330–425
475. Someren RNM,van, Coleshaw SRK, Mincer PJ, Keatinge WR (1982) Restoration of thermoregulatory response to body cooling by cooling hand and feet. J Appl Physiol 53: 1228–1233
476. Somero GN (1997) Temperature relationships: from molecules to biogeography. In: Dantzler WH (ed) Handbook of physiology, sect 4: Comparative physiology, vol II. Oxford University Press, New York, pp 1391–1444
477. Spaan G, Klußmann FW (1970) Die Frequenz des Kältezitterns bei Tierarten verschiedener Größe. Pflügers Arch 320: 318–333
478. Stelzner JK, Hausfater G (1986) Posture, microclimate, and thermoregulation in yellow baboons. Primates 27: 449–463
479. Stephenson LA, Wenger CB, Odonovan BH, Nadel ER (1984) Circadian rhythm in sweating and cutaneous blood flow. Am J Physiol 246: R321–R324
480. Stitt JT (1976) The regulation of respiratory evaporative heat loss in rabbit. J Physiol (Lond) 258: 157–171
481. Stitt JT (1980) Variable open–loop gain in the control of thermogenesis in cold–exposed rabbits. J Appl Physiol 48: 495–499
482. Stitt JT (1986) Prostaglandin E as the neural mediator of the febrile response. Yale J Biol Med 59: 137–149
483. Stitt JT (1991) Differential sensitivity in the sites of the fever production by prostaglandin E1 within the hypothalamus of the rat. J Physiol (Lond) 432: 99–110
484. Stitt JT, Adair ER, Nadel ER, Stolwijk JAJ (1971) The relation between behavior and physiology in the thermoregulatory response of the squirrel monkey. J Physiol (Paris) 63: 424–427
485. Stitt JT, Hardy JD, Stolwijk JAJ (1975) PGE1 fever: its effect on thermoregulation at different low ambient temperatures. Am J Physiol 227: 622–629
486. Stolwijk JAJ, Nadel ER (1973) Thermoregulation during positive and negative work exercise. Fed Proc 32: 1607–1613
487. Takamata A, Mack GW, Gillen CM, Nadel ER (1994) Sodium appetite, thirst, and body fluid regulation in humans during rehydration without sodium replacement. Am J Physiol 266: R1493–R1502
488. Takamata A, Nagashima K, Nose H, Morimoto T (1997) Osmoregulatory inhibition of thermally induced cutaneous vasodilation in passively heated humans. Am J Physiol 273: R197–R204
489. Taylor CA (1991) Surgical hypothermia. In: Schönbaum E, Lomax P (eds) Thermoregulation: pathology, pharmocology, and therapy. Pergamon Press, New York, pp 363–396
490. Taylor CR (1970) Dehydration and heat: effects on temperature regulation of East African ungulates. Am J Physiol 219: 1136–1139
491. Taylor CR (1977) Exercise and environmental heat loads: different mechanisms for solving different problems? In: Robertshaw D (ed) International review of physiology, vol 15. Environmental physiology II. University Park Press, Baltimore, pp 119–146
492. Taylor CR (1994) Relating mechanics and energetics during exercise. Advan Vet Sci Comp Med 38A: 181–215
493. Taylor CR, Lyman CP (1972) Heat storage in running antelopes: independence of brain and body temperatures. Am J Physiol 222: 114–117
494. Taylor CR, Maloiy GMO, Weibel ER, Langman VA, Kamau JMZ, Seeherman HJ, Heglund NC (1981) Design of the mammalian respiratory system. III. Scaling maximum aerobic capacity to body mass: wild and domestic mammals. Respir Physiol 44: 25–37
495. Taylor CR, Rowntree VJ (1973) Temperature regulation and heat balance in cheetahs: a strategy for sprinters? Am J Physiol 224: 848–853

496. Taylor CR, Rowntree VJ (1974) Panting vs. sweating: optimal strategies for dissipating exercise and environmental heat loads. Proc Int Union Physiol Sci 11: 348 (Abstract)

497. Taylor CR, Schmidt–Nielsen K, Dmi'el R, Fedak M (1971) Effect of hyperthermia on heat balance during running in the African hunting dog. Am J Physiol 220: 823–827

498. Taylor WF, Johnson JM, O'Leary DS, Park MK (1984) Effect of high local temperature on reflex cutaneous vasodilation. J Appl Physiol 57: 191–196

499. Templeton JR (1970) Reptiles. In: Whittow GC (ed) Comparative physiology of thermoregulation. Academic Press, New York, pp 167–221

500. Thauer R (1939) Der Mechanismus der Wärmeregulation. Ergebn Physiol 41: 607–805

501. Thauer R (1965) The circulation in hypothermia of nonhibernating animals and men. In: Hamilton WF, Dow P (eds) Handbook of physiology, sect 2, vol III, Circulation. American Physiological Society, Washington, D.C., pp 1899–1920

502. Thauer R (1965) Circulatory adjustments to climatic requirements. In: Hamilton WF, Dow P (eds) Handbook of Physiology, sect 2, vol III, Circulation. American Physiological Society, Washington, D.C., pp 1921–1965

503. Thermal Commission (1987) Glossary of terms for thermal physiology. Pflügers Arch 410: 567–587

504. Thornhill JA, Halvorson I (1994) Electrical stimulation of the posterior and ventromedial hypothalamic nuclei causes specific activation of shivering and nonshivering thermogenesis. Can J Physiol Pharmacol 72: 89–96

505. Timbal J, Boutelier C, Loncle M, Bougues L (1976) Comparison of shivering in man exposed to cold in water and air. Pflügers Arch 365: 243–248

506. Toner MM, McArdle WD (1996) Human thermoregulatory responses to acute cold stress with special reference to water immersion. In: Fregly MJ, Blatteis CM (eds) Handbook of physiology, sect 4: Environmental physiology, vol I. Oxford University Press, New York, pp 379–397

507. Trayhurn P, Nicholls DG (1986) Brown adipose tissue. Edward Arnold, London, pp 1–374

508. Turek FW, Van Reeth O (1996) Circadian rhythms. In: Fregly MJ, Blatteis CM (eds) Handbook of physiology, sect 4: Environmental physiology, vol II. Oxford University Press, New York, pp 1329–1359

509. Underwood CR, Ward EJ (1966) The solar radiation area of man. Ergonomics 9: 155–168

510. Van Zoeren JG, Stricker EM (1977) Effects of preoptic, lateral hypothalamic, or dopamine–depleting lesions on behavioral thermoregulation in rats exposed to the cold. J Comp Physiol Psychol 91: 989–999

511. Vane JR (1971) Inhibition of prostaglandin synthesis as a mechanism of action for aspirin–like drugs. Nature New Biol 231: 232–235

512. Vanhoutte PM (1983) Physical factors of regulation. In: Shepherd JT, Abboud FM, Geiger SR (eds) Handbook of physiology, sect 2, vol II: Vascular smooth muscle. American Physiological Society, Bethesda, MD, pp 443–474

513. Waites GMH (1962) The effect of heating the scrotum of the ram on respiration and body temperature. Q J Exp Physiol 47: 314–323

514. Walther OE, Simon E, Jessen C (1971) Thermoregulatory adjustments of skin blood flow in chronically spinalized dogs. Pflügers Arch 322: 323–335

515. Wang LCH (1973) Radiotelemetric study of hibernation under natural and laboratory conditions. Am J Physiol 224: 673–677

516. Wang LCH, Lee TF (1996) Torpor and hibernation in mammals: metabolic, physiological, and biochemical adaptations. In: Fregly MJ, Blatteis CM (eds) Handbook of physiology, sect 4: Environmental physiology, vol I. Oxford University Press, New York, pp 507–532

517. Wathes CM, Jones CDR, Webster AJF (1983) Ventilation, air hygiene and animal health. Vet Rec 113: 554–559

518. Watkins LR, Maier SF, Goehler LE (1995) Cytokine–to–brain communication: a review and analysis of alternative mechanisms. Life Sci 57: 1011–1026

519. Watts PD, Oeritsland NA, Jonkel C, Ronald K (1981) Mammalian hibernation and the oxygen consumption of a denning black bear (*Ursus americanus*). Comp Biochem Physiol (A) 69: 121–123

520. Webb P (1992) Temperatures of skin, subcutaneous tissue, muscle and core in resting men in cold, comfortable and hot conditions. Eur J Appl Physiol 64: 471–476

521. Webb P (1993) Heat storage and body temperature during cooling and rewarming. Eur J Appl Physiol 66: 18–24
522. Webster AJF (1966) The establishment of thermal equilibrium in sheep exposed to cold environments. Res Vet Sci 7: 454–465
523. Webster AJF (1974) Adaptation to cold. In: Robertshaw D (ed) Environmental physiology, MTP International review of science, Physiology Series One, vol 7. Butterworths, London, pp 71–106
524. Webster AJF, Chlumecky J, Young BA (1970) Effects of cold environments on the energy exchanges of young beef cattle. Can J Anim Sci 50: 89–100
525. Webster AJF, Hicks AM, Hays FL (1969) Cold climate and cold temperature induced changes in the heat production and thermal insulation of sheep. Can J Physiol Pharmacol 47: 553–562
526. Weibel ER, Taylor CR, Hoppeler H (1992) Variations in function and design: Testing symmorphosis in the respiratory system. Respir Physiol 87: 325–348
527. Welch WJ (1992) Mammalian stress response: cell physiology, structure/function of stress proteins, and implications for medicine and disease. Physiol Rev 72: 1063–1081
528. Wenger CB, Roberts MF (1980) Control of forearm venous volume during exercise and body heating. J Appl Physiol 48: 114–119
529. Wenzel HG, Piekarski C (1984) Klima und Arbeit. Bayerisches Staatsministerium für Arbeit und Sozialordnung, München, pp 1–200
530. Werner J (1996) Modeling homeostatic responses to heat and cold. In: Fregly MJ, Blatteis CM (eds) Handbook of physiology, sect 4: Environmental physiology, vol I. Oxford University Press, New York, pp 613–626
531. Werner J, Buse M (1988) Temperature profiles with respect to inhomogeneity and geometry of the human body. J Appl Physiol 65: 1110–1118
532. Wit A, Wang SC (1968) Effects of increasing ambient temperature on unit activity in the preoptic–anterior hypothalamus (PO/AH). Am J Physiol 215: 1151–1159
533. Withers PC (1992) Comparative animal physiology. Harcourt Brace Jovanovich College Publishers, Forth Worth, pp 1–949
534. Wyndham CH, Strydom NB, Morrison JF, Williams CG, Bredell GAG, Maritz JS, Munro A (1965) Criteria for physiological limits for work in heat. J Appl Physiol 20: 37–45
535. Wyndham CH, Williams CG, Loots H (1968) Reactions to cold. J Appl Physiol 24: 282–287
536. Wyndham CH, Williams CG, Morrison JF, Heyns AJA, Siebert J (1968) Tolerance of very hot humid environments by highly acclimatized bantu at rest. Br J Ind Med 25: 22–39
537. Wyss CR, Brengelmann GL, Johnson JM, Rowell LB, Niederberger M (1974) Control of skin blood flow, sweating and heart rate: Role of skin versus core temperature. J Appl Physiol 36: 726–734
538. Yahav S, Buffenstein R (1991) Huddling behavior facilitates homeothermy in the naked mole rat *Heterocephalus glaber*. Physiol Zool 64: 871–884
539. Yang RC, Mack GW, Wolfe RR, Nadel ER (1998) Albumin synthesis after intense intermittent exercise in human subjects. J Appl Physiol 84: 584–592
540. Young AJ (1996) Homeostatic responses to prolonged cold exposure: human cold acclimatization. In: Fregly MJ, Blatteis CM (eds) Handbook of physiology, sect 4: Environmental physiology, vol I. Oxford University Press, New York, pp 419–438
541. Young DR, Mosher R, Erve P, Spector H (1959) Energy metabolism and gas exchange during treadmill running in dogs. J Appl Physiol 14: 834–838
542. Yousef MK (1985) Thermoneutral zone. In: Yousef MK (ed) Stress physiology in livestock. vol I: Basic principles. CRC Press, Boca Raton, pp 67–74
543. Zeisberger E (1999) From humoral fever to neuroimmunological control of fever. J Therm Biol 24: 287–326
544. Zitnick RS, Ambrosioni E, Shepherd JT (1971) Effect of temperature on cutaneous venomotor reflexes in man. J Appl Physiol 31: 507–512

Index